Other books by Rochester Veterans Writing Group:

- ***United in Service * United in Sacrifice***, 2020
- ***...NOT Fade Away***, 2023

The Lighter Side of Military Life

RVWG

Rochester Veterans Writing Group

We dedicate this collection of military stories to Chuck (Charles F. Willard Jr.) who we lost on May 14, 2024. He was an "amazing, kind gentleman" who shared his life, love of writing, many experiences, and editing skills with us. We honor his final wish to rejoice in the fact that our paths crossed.

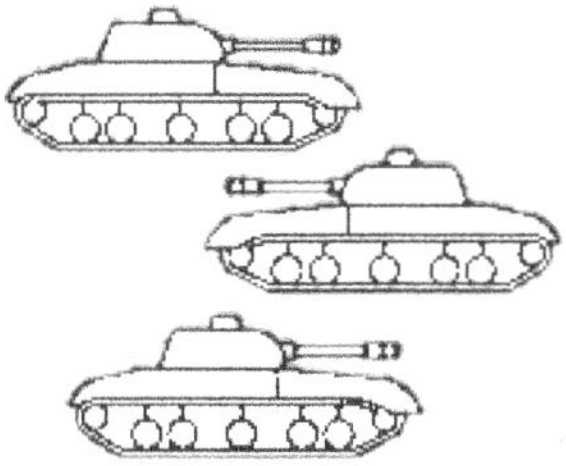

The Lighter Side Of Military Life

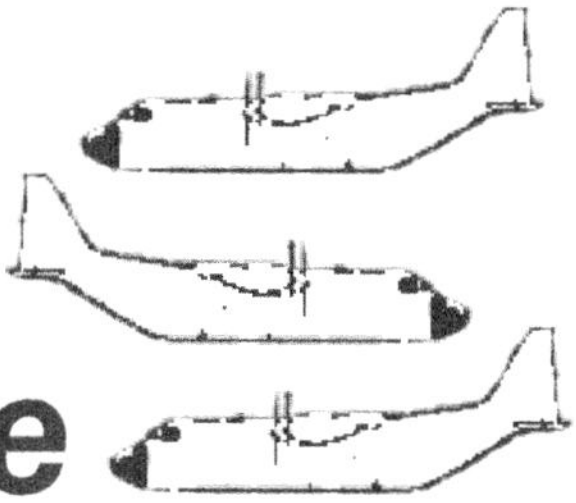

Military Humor

from the

Rochester Veterans Writing Group

And Friends

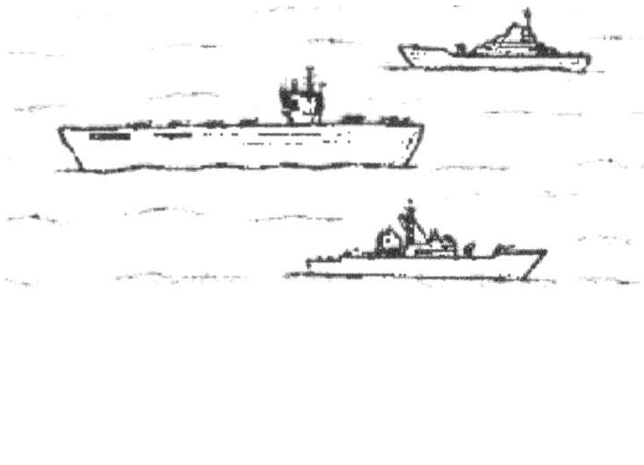

The Lighter Side of Military Life

Rochester Veterans Writing Group

Rochester, New York

Stories "No PhD Needed," and "Cat Nipped" were previously published in *Afterwords*©, an anthology from the Veterans Writing Workshop 2020 Hudson, NY, by Cindy Blair

Stories "Private Skibootch," "Private First Class Skibootch," and "Dienstglas Skibootch" were previously published in *O'Book*, by Robert M. Whelan, ©2023

ISBN 978-0-9700441-4-3

Edited by Green Shadow Press

Lead pages and P. 248 image by Gary Redlinski

Cover Art by Christopher Cilento of Ravenworks Studios

TABLE OF CONTENTS

"WAR & TRAINING IS HELL" 1

Commando Training---3

Joe Mele

At Ease Troop—Smile If You've Got One!!! -------------------- 18

Richard Terboss

This Is My First Rodeo -- 23

John Steele

Lighter Side of Offloading a C-130 at Lunch Time------------ 31

Gary Redlinski

The Fly By -- 33

Arash Rivazfar

Little Stories --- 34

John Michael (Mike) Ryan

Second Amendment Adventures—A Tale of Four Chapters

The only good part: It wasn't raining. Chapter 1 --------------- 37

Vaughn Stelzenmuller

FIRST THINGS FIRST: FAMILY! 41

Always Gonna Be a Country Girl ------------------------------- 43

Sue Spitulnik

One of Those Days --- 45

Vaughn Stelzenmuller

The Screaming Summer of 2003 ------------------------------- 47

Lee Nelson

See Ya Later ---------- 57

Charles F. Willard

I Lost My Son ---------- 60

Sue Spitulnik

U.S. Army Story ---------- 63

Richard DeLisle

The Bomber Suit ---------- 69

Charles F. Willard

The Disappearing Buddha ---------- 74

Richard Terboss

11 General Orders for Sea (Service) Story ---------- 75

Noel Nichols

The Pastor Who Wore Khakis ---------- 76

Father William B. Leone

WHAT COULD GO WRONG? 79

What Could Go Wrong? A Trilogy ---------- 81

Dave Cole

Battalion Surgeon Journal Winter 1945 ---------- 88

Joe Mele

Invader ---------- 94

Janice Bilohlavek

Toys in the Army ---------- 95

John Steele

Carry a Gun in Nowhere ---------- 103

Lee Nelson

Some Navy Memories---108
John Michael (Mike) Ryan

Hurry Up and Wait---111
Sue Spitulnik

Beer Adventure in Saigon---115
Gary Redlinski

The Specialist---117
Tim Hansen

An Epicurean Disaster---119
Chris Cilento

Do Not Put Your Finger In This Crack---128
John Michael (Mike) Ryan

Henry V^th Addresses the Troops. Chapter 2---130
Vaughn Stelzenmuller

ENLISTED 2, OFFICER 1 ... 133

Humor of an Army Mortuary Identification Specialist---135
Gary Redlinski

At the Stroke of God-O'clock---137
Vaughn Stelzenmuller

No PhD Needed---138
Cindy Blair

Officers---140
John Steele

Slow Night at the Mortuary---145
Gary Redlinski

General Electric Issues Orders ---------------------------------147
Dana Hill

On Parade ---148
Vaughn Stelzenmuller

DON'T SCREW THIS ONE UP...149

The Shackleton, the Cockney, and the Yank -----------------151
Charles F. Willard

Bake Sale Saturday---154
Vaughn Stelzenmuller

Driving in Florida --156
Lee Nelson

MILES Gear --157
John Steele

Private Skibootch --161
Bob Whelan

Private First Class Skibootch ----------------------------------164
Bob Whelan

Dienstglas Skibootch ---167
Bob Whelan

David Niven Forms an Army. Chapter 3---------------------169
Vaughn Stelzenmuller

BETTER LUCKY THAN GOOD...173

Initial Operating Experience ----------------------------------175
John P. Bland

Ft Belvoir Diary ---180
Richard DeLisle

The Upside-Down Spin -- 183
Charles F. Willard

Blessèd Toys -- 186
Vaughn Stelzenmuller

Collision Course --- 188
Charles F. Willard

Down Time Ditherings --- 192
Richard Terboss

DANGER! CELEBRITY ON BOARD! 195

Live From Kabul! --- 197
Tim Hansen

Farmer --- 201
Charles F. Willard

Glenn Ford and the Ouija Board ------------------------------- 203
John P. Bland

Cat Nipped --- 206
Cindy Blair

Toby Keith -- 208
John Steele

"HANDS ACROSS THE SEA" (OR COUNTRY) 215

Marlboro Country -- 217
Charles F. Willard

Are You There, Moriarty? -- 219
Vaughn Stelzenmuller

Laughable Instances --- 221
Sue Spitulnik

Starter Home --226
Lee Nelson

LD's Value--233
John P. Bland

Made in Korea--236
Tim Hansen

Heterogeneous Recruits at Basic Training --------------------239
John Steele

Beware of the Seabat --244
John Michael (Mike) Ryan

The Light at the End of the Gun Barrel. Chapter 4 ------------246
Vaughn Stelzenmuller

LAST CHAPTER: UNCHAINED MILITARY HUMOR...............249

Hot Dogs --251
Lee Nelson

Being Abandoned on Thanksgiving ---------------------------254
John Steele

CHUCK WILLARD TRIBUTES 259

AUTHORS...277

"WAR & TRAINING IS HELL"

COMMANDO TRAINING

By Joe Mele

"War is hell" began his war stories. Few and far between, his accounts were all preceded by this three word editorial warning and philosophical truth about the nature of war. My sisters and I were awed and scared as our father's impassioned monologues (sanitized for our young ears) provided brief glimpses into combat in the European Theater of War (World War II). As the surgeon for the 749th Tank Battalion, he had witnessed up close the gruesome toll that bombs and bullets took on flesh and bone.

More than a decade removed from the war, the grim ghosts loosed their grip on his psyche, and he remembered and laughed about lighter moments of his military service. The nascent medium of television helped in this regard. We'd gather as a family in our small den, all eyes on the seventeen inch screen, to watch Sergeant Bilko (Phil Silvers) fast talk and scheme his way into and out of predicaments with his staff of motor pool screw-ups.

My father didn't preface the following story with "War is Hell." "Training is Hell" would have worked. He didn't say that either. Although this event occurred while he was still with the 747th Tank Battalion, I suspect his memory was jogged by a reunion of the 749th he attended a few weeks before he told me the tale.

Sitting in a lawn chair under the shade of a mature backyard maple tree, he sipped a dry martini (gin and a splash of dry vermouth over ice with a lemon twist), as he waited for the charcoal in the portable grill to glow red hot. I was nearby, flopping in the grass trying to nail a handstand. His spirits lifted, and his tongue loosened by the pre-dinner cocktail, he asked me "Joe, did I tell you about the commando training in Texas?" "No" I replied—and meant it. I heard many of his stories more than once, but not the war stories. He never repeated those.

Here it is, knitted together the best I can from the bits and pieces I recall about that conversation sixty years ago.

During U. S. Army Commando Training at Camp Hood, TX, he was responsible for providing medical care to the participating soldiers. He set up first aid stations at key points along the route, and rotated his medics among them. One aspect of the training was a five-mile slog, required of everyone, including officers. He was slim, but hadn't yet developed the endurance demanded of a long run—his work requiring and his inclination leaning toward more intellectual pursuits. And medical officer's training was less physically rigorous than other enlistee's.

As he was trudging, sweating, and panting along the dusty Texas road, he flagged down a passing truck and hitched a ride. "*Salvation*," he thought as he hopped into the passenger seat. Even before the driver could shift into first gear, a gravelly voice bellowed through the open passenger side window "Not so fast. Everybody runs." A Captain overseeing this section of the course had spotted him. With all the sincere conviction he could muster Lt. Mele replied, "Sir, I can run, or I can manage the First Aid Stations—can't do both. Its impossible to do both. There are three first aid stations, an ambulance, and eight medics to supervise. Guaranteed there will be casualties during commando training. I'm the only doctor in this outfit. Who else is gonna stitch up lacerations? Can't do both sir." (I imagined him pleading his case in the rapid fire style of Sgt. Bilko.) The Captain's stone face revealed nothing, and Lt. Mele suspected he wasn't buying it, but the Captain did not press the issue, and, Lt. Mele got away with one.

Decades later, among the letters, papers, and documents that my mother saved, I found two items related to this event. One was a commendation for his initiative, ability, and ingenuity in deploying his men as they provided first aid to participants in the commando training. The other document is the March 28, 1943 edition of TANK TALK. Both are below.

50

HEADQUARTERS
ADVANCED UNIT TRAINING CENTER
TANK DESTROYER CENTER
CAMP HOOD, TEXAS

IN REPLY REFER TO:

March 22, 1943

Subject: Commendation of 1st Lieutenant Joseph Mele, Medical Detachment.

To : Commanding Officer, 747th Tank Battalion, Camp Hood, Texas.

1. During the week the 747th Tank Battalion was undergoing training on the Tank Hunting Course, 1st Lieutenant Mele, in charge of the Medical Detachment, took advantage of the opportunity to train his enlisted men in attending to their duties under actual field conditions. He did not establish the usual medical set-up in the building provided, but established a field set-up near the Infiltration Course where he remained at all times except when needed for emergencies. He kept his ambulance and driver, properly equipped, ready for instant use on call. He stationed one of his men, properly equipped, at the Woods Fighting Course; one at the Street Fighting Course; and one at the Infiltration Course at all times. During the time the Battalion was negotiating the Obstacle Course he stationed his enlisted personnel at each danger point, and himself stayed at the most dangerous obstacle with the ambulance.

2. He is to be commended for his foresight in planning and carrying out these methods without instructions from the Tank Hunting Course Commander. He showed initiative, ability and ingenuity in these arrangements in providing field training for his men and so rotating these men that each had an opportunity to participate in the actual training being given to the other members of the Battalion.

Ralph W. Sleator

RALPH W. SLEATOR,
Major, Infantry,
Ins.-Ins., Tank Hunting.

Lt Joseph M Mele

TANK TALK

Vol. 1 | COMMANDO ISSUE | No. 2

COMMANDO COURSE COVERED

MEDICAL DETACHMENT OPERATES WITH SKILL AND DISPATCH

Lt. Mele's "pillrollers" were complimented freely on the excellent manner in which they performed their duties on the recent laceration marathon carried out on the fatal number three Commando Course.

Maj. Sleator, inspector instructor of the course, remarked thatour medics were the best posted crew he had seen to date.

Exact statisticsare lacking but 1st. Sgt. Filinuk says that somewhere in the neighbor-hood of 11 barrels of iodine were used, (some neighborhood). At any rate ALL the adhesive tape was expended (more was sent for) much of it however, in the name of modesty; the barbed wire barricade making exposed posteriors the order of the day.

A thousand and some cuts, scratches, and bruises were treated. One operation performed on

(cont'd. page10, col. 2)

Don't bother looking for IT. See page five.

"BROKEN BONE VALLEY" CONQUERED BY 747TH.

The Battle of Waterloo, the Meuse-Argonne Offensive and the Evacuation of Dunkirk all rolled into one would furnish the only basis upon which this reporter could compare the recent activities of our

MARCH 14th. MARCH 20th.

battalion on that Satanic Safari, that camouflaged Medusa's Head, that G.I. Hall of Horrors known as Commando Course #3.

The 747th. Tank Bn. left here on Sunday afternoon, March 14th. to bivouac in the area near Commando Town for six days of intensive training in commando tactics. Having removed their swaddling clothes (basic training is over) the battalion was due for a little toughening up.... little? Little did they know!

The training schedule included instruction in the following: Close Combat, Woods Fighting, Battle Firing, Night Fir-

(continued on page three)

THE SPIRIT OF THE 747TH.

Pvt. Paul Liebowitz of Co. C lost his glasses prior to crawling the infiltration course at "Commando Town" last week.

"Hey Sarge" he said to S/Sgt. Pemata, "tell me where the hell the shell holes, dynamite and barbed wire are and I'll follow you through."

With live ammunition whistling 20 inches over his head, and grenades and dynamite blowing up around him this private crawled the entire course asking a direction now and then from his sergeant AND HE DIDN'T FINISH LAST!

BUY A WAR BOND ON PAY DAY

TANK TALK
Page 2
FIVE MILE RUN
CAPT.
TNT
LT. JORDAN
LT. BOARDMAN
LT. GROSS
LT. SCOTT
GUESS WHO
COMMANDO COURSE #3
CAPT. SEBESTA
LT. BIGELOW
MAJ. FRIES
ZIPKIN

COMMANDO COURSE COVERED
(continued from page one)

ing, Scouting and Patrolling, Infiltration, Field Craft, Street Fighting, and Demolitions. In the last named we were taught the use of "Mollies". Incidentally all you need is a sock, a handful of grease, and a little gunpowder and boy oh boy what you can do to a tank. Holy Mackerel! Why are WE getting so enthusiastic?

"Choke me," said Sgt. Zidelsky, the Judo instructor in Close Combat. So we did. Jerk.....Swoosh.....WHAM! "See" he'd say, bending over our broken body, "Nuttin' to it, nuttin' to it" (south Brooklyneese), and believe us there was plenty of "nuttin' to it." In this delicate manner the entire battalion learned about the gentle art of murderous assault with intent to maim.

Ah, but we left one out, the OBSTACLE CURSE! (That's no misprint) Lt. Col. Kimbrell, "father" of the course, has fought with the British commandoes. He participated in two British raids on the French Coast. We would like to point out to the Colonel that Nero was a novice! His Commando Course is a masterpiece of unrefined torture.

Take that second wall for instance, you know the fifteen foot casualty list. That's where they begin. They break your arches first just to make it a little harder. Next comes the barbed wire field that has to be crawled thru on your back with the damn stuff tearing T-Bone steaks out of your short loins while some - - - - keeps throwing live grenades between your feet and another - - - - waves a stinking smoke pot in your face and yells, "get going" (good advice but entirely unecessary).

Then you get to the rope climb, twenty feet of hand over hand to a platform, and down the other side the same way. Of course by then you've only torn the flesh off both palms and your biceps feel like shredded wheat.

Do they stop you then. Hell no! Right in front of your eyes and straight up till your neck hurts is a natural montrosity called "Puke Hill." You're ready to die peacefully and quietly and they yell, "Double Time Up That Hill." The sweetish smell of dynamite fumes and smoke pots goes down your throat, reconnoiters and then comes back up with reenforcements. It certainly brings out the best in you!

After that, a little thing like the infiltration course was a snap. They merely ask you to crawl on your belly "woim-like" thru barbed wire, over low mounds and around shell holes, meanwhile touching off a stick of dynamite here and there for realism. A .30 calibre machine gun plays, "Come All Ye Faithful" 30 inches above the whole thing which is 20 to 24 inches above you, depending on the size of your underwear.

Street fighting was, so to speak, right up our alley. Climbing a rooftop with a rifle in one hand is really a difficult thing to do. Targets pop up here and there generally at a difficult angle and always when you are
(continued on page 11)

TANK TALK

TANK TALK is published bi-weekly by the enlisted men of the 747th. Tank Battalion (M), Camp Hood, Texas.

Capt. C.A.Jordan............Supervisor

Vol. 1 MARCH 28,1943 No. 2

Editors

Pvt. RAY GREENBERG Pvt. RAY LESSER

Art

Pvt. HAROLD ZIPKIN

THE SPIRIT OF THE 747th.

On page one of this issue appears the story about Pvt. Paul Liebowitz who; despite the loss of his glasses, crawled through gunfire to complete the infiltration course on the Tank Hunting Range.

Pvt. Liebowitz could have very easily "goldbricked" his way out of the course but he had the courage and the spirit that goes to make up the fighting man that is going to help win this war.

There are many men who are constantly looking for ways to "goldbrick" out of various activities; they are hurting only themselves for it is this vital training that may some day save their life.

It is men like Pvt. Liebowitz that will make the 747th. a coordinated fighting outfit capable of undertaking any task assigned them.

Are YOU another Pvt. Liebowitz or just a "goldbrick?" HAVE YOU GOT THE SPIRIT OF THE 747th.?

NOTICE

There are still many openings for men wishing to write for TANK TALK, or for men with typing ability. If interested contact Ray Greenberg of his company, (Service Co.)

by PVT. RAY P. LESSER

A couple of steel "ducks" flew in a few days ago. Those squat chunky looking things are amphibian peeps. The Recon gang still looks a little bewildered behind the wheel but they'll get used to them.. Now if they'll only ~~put wings on those peeps we'll even be~~ able to do our own aerial observation.

T/Sgt. Carver is all burnt up. "Listen," he warns, "keep those blankety blank tanks off those black top roads, the steel tracks on those babies will tear hell out of the surface. A sharp turn with rubber tracks will do it also. If you want to stay off of Road Repair details, heed or bleed.

You can save time and trouble says T/Sgt. Givens of A Co., if you check your fuel systems in the M 4's and M 3's more often. Gas tank rust is the enemy to watch for.

There must be a lot of orstrich blood in some of these soldiers. We noticed how carefully many of them camouflaged their tanks and vehicles (on the recent bivouac) and then went out and pitched their tents where they didn't even have protection from a low flying pigeon. The sloping side of a tent looks like a nice new shiny dime when viewed from the air on a sunny day. We know because we've seen them. COVER UP!

WE THANK YOU ---

We, the staff of TANK TALK thank you all for the manner in which you all received the first issue.

We hope that you will derive thesame pleasure from future issues as we will have in printing them.

What is this rumor about a beauty contest for the wives and sweethearts of our battalion. Seems like someone has an idea that the winner of said contest could be brought here, christen a tank with her name on it and her lucky sponsor would be given a three day pass. Picture pickers will be from another outfit. So much per picture would build the bankroll for transportation, hotel, etc. What do you think about it? Let's hear from you.

800 (and some) pairs of coverall's will be salvaged as a result of the recent epidemic of broken-field crawling.

It sure did our heart good. We mean that operation performed by Lt. Mele on that commando instructor. He took a piece of steel out of this Sgt's. backside without the benefit of anesthetic. Ha! Ha! How do YOU like it Sgt? .....The medics used so much adhesive tape on Sgt. "Big Red" DeSanti that he began to look like a badly worn inner tube.

Pvt. Moscato surprised everyone on that infiltration course. Nice goin' bud.......Rumor has it that T/Sgt. Cox and PFC Day of C Co. looked like one man going thru that "no man's land."

We'd like to know why everybody's posing for pictures around here. It's getting to look like Hollywood Blvd. What's gonna be?

A reliable source informs us that Capt. Nelson says he will never run with Lt. Munzer again. Five miles in 52 minutes. Who can blame him?

"Sgt. give me your coveralls" and 1st. Sgt. Nickell of Service Co. obliged the Major by removing them so that the Major could use them on the windy infiltration course. This officer (not a member of the 747th) did very well. Sgt. Nickell froze his.....feet off.

Speaking of first soldiers, we saw it with our own eyes, and we're now willing to die, having seen everything. Yes, sir, 1st. Sgt. Filimuk of the medics was so short handed the other day that we caught him (red handed) WASHING A PEEP! Honestly, what is the army coming to?

Service Company had a whale of a battle the other night; it turned out to be shadow boxing. What they thought was a commando raid on their area turned out to be two woodpeckers trying to eat a G.I. biscuit. What do you know about it Sgt. Frisch?

Alibis keep coming in on the record firing. Let all those with complaints please lodge them with the T.S. department; they will be consolidated and reported to the Chaplain.

(continued on next page)

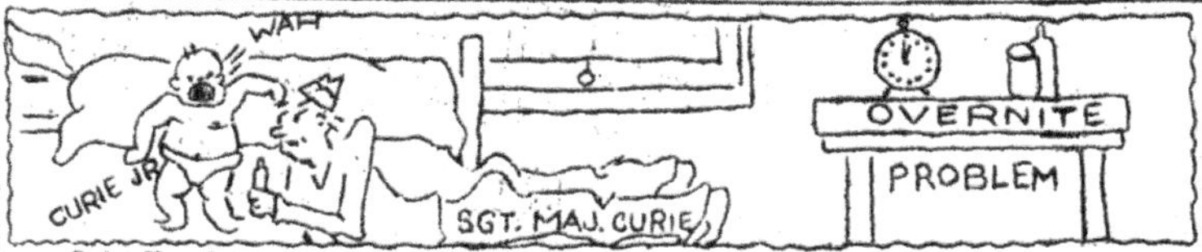

Sgt. Van Pelt, Pvt. Lesser and Pvt. Psamodikis of Hq. Co. seem fond of knocking down trees with flying tackles; where do you buy the stuff fellas?

Lts. Halpin, Cantrell, and Gross of Hq. Co. joined their EM in a football game during "Hell Week." The game was played on a field that looked like it had been thru the siege of Stalingrad. Score: 11 outs, 10 nicks, and no terrors.

CLASSIFIED AD: S/Sgt. Ulrich of Hq. Co. wants to rent a house somewhere near. He days last time he came home his little daughter shrieked, "Look Ma, that soldiers here again."

They say, and we can't verify it that once upon a time a certain 1st. Sgt. (initials J.B.) upon learning of a fatal accident to a member of the company wrote (after the man's name) in the morning report, "Duty to Death." Upon discovering the next day that the man was only slighty injured, he made the following entry, "Pvt. John Doe, Death to Duty."

COMMANDO CAPERS: While on the Commando Course, who were the fellows that pitched their tents almost directly over and old latrine?.....Demou's hunger isn't satisfied until he can travel through that chow line at least three times, even though he gets served only once. Isn't that right Geronimo? ...All the "goldbricks" that volunteered for K.P. in the field are getting their share of it here in camp.......I wonder, did J.J. see that tracer shell that had his serial number on it? He must have out-maneuvered it while on the infiltration course......."Greenberg, the Irishman".......Rapp was on the verge of having his most learned colleagues practice their trade on him when he decided to try out for a pair of wings. That wall was no place for an undertaker to take a spill, soldier. There's too many fellows that know how to dig a 6x6x3.......To all the short guys that couldn't make the first wall, "Why Don't Ya Grow Up?".......Bulnes, how's your neck after that close-combat sergeant got through with you?.......The hero of the month is Ace Respoli of Service Co. who, while out on the Commando Course killed a copperhead.......whose girl in Waco asked him, "Please don't get killed on the Commando Course".......Imagine the expression on the faces of the two privates thumbing their way out to the course from camp one NIGHT when they received a lift from their own C.O........ Who broke the record for the slowest time on the five mile run, yet has the longest pair of legs and biggest pair of feet in the battalion (sorry D.S.).... Service Co. has another Cunningham; he ran the five miles in 47 minutes.....The best remedy for a case of sore eardrums is a week on K.P. Ask C.M. - he'll tell you.......Leg motion is walking again.......Master Sgt. Kirkendall...Quote, "I can drive Sarge - only I get nervous." - G.D......."My Darling Husband" wasn't such a darling not so long ago.......Say Nick, isn't Service Co. very observant about quarantines?.......What private in Service Co. recently received a telephone bill for $61 which had the following note, "Happy Birthday."

747TH STATISTICS

Congratulations to: J. Adesman, 128; M. Berman, 143; L. Blum, 113; R. DiClemente, 134; S. Dunefsky, 120; I. Frankel, 120; P. Gillen, 132; N. Hanover, 121; T. Hensom, 127; R. Hompesch, 135; S. Kagen, 119; V. Morales, 116; H. Pauchey, 129; J. Pressler, 114; M. Rosenfeld, 110; Wm. Shann, 125; B. Sherman, 162; A. Steinman, 117; and W. Weiss, 110 who passed the army Specialized Training Test given here Saturday morning, March 13th., and who are now at school.

Those men wishing to know their score on this test should see Sgt. Quist at battalion headquarters.

* * * * *

The 747th. can boast among its men Pvt. John Rosselet, Vice-President of the National Horseshoe Pitchers Assn., and who is a nationally-ranked pitcher. Pvt. Rosselet has been given permission to set up his court on the area, and will soon give an exhibition of trick horseshoe pitching. Pvt. Rosselet also challanges any and all men in the battalion who think they can out-pitch him.

* * * * *

Omitted from the last issue of TANK TALK were Cpl. Milwood, 96; Lt. Hardy, 90; and P.F.C. H. Pauchey, 38; who made the three highest scores for C Co. for record firing of the Thompson Sub-Machine Gun.

* * * * *

Statistics show that over 10 percent of the battalion is now away at schools.

MERCY
OH! OH! OH! OH!

We love THE GUY who chases us poor buck privates all over the battalion area every Friday night.

Guess Who?

"I want 4 volunteers, - - - YOU, YOU, YOU, and YOU!

COMMANDO COURSE NUMBER THREE

I wish a certain Cpl. in service company was C.Q. more often.

When are we going to get those foot lockers.......SIX MONTHS AFTER?

TANK TALK IS YOUR PAPER! WHAT DO YOU SAY FELLOWS - LET'S HEAR FROM YOU. IF YOU HAVE ANY PET PEEVES, GRIPES OR GROANS SEND THEM IN TO US FOR PUBLICATION, OR FOREVER HOLD YOUR PEACE!

ZIPKIN DOESN'T LIKE THIS COMMANDO BUSINESS! IT GIVES HIM OBSTACLE ILLUSIONS.

Oh, what a lovely state Texas is, to STAY AWAY FROM.

We love these self-styled barrack Napoleons who supervise the evening blackouts at 9 p.m. sharp.

Are officers and N.C.O.'s immune to the Measles???

OH! FOR THE LIFE OF A NEWSPAPERMAN

That's all for this issue fellows, but let's keep hearing from you - or do you really love Texas?

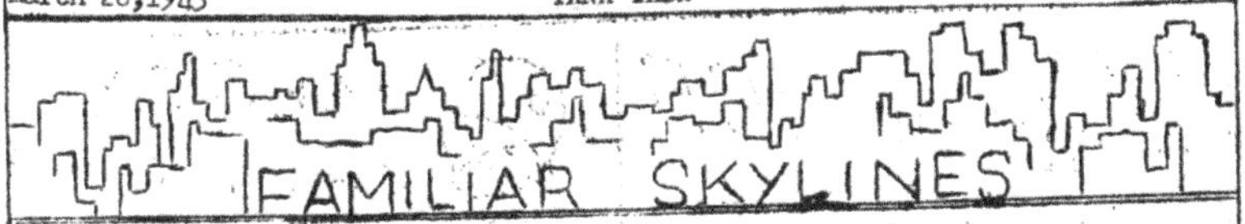

Very few people in New York or that TOWN, Jersey have seen many tank emblems - so fellows start making up a line and prepare for the questions you will probably be asked.

Brooklyn will be in the lead again this year, but how about the Giants?

Spring will soon touch the North and every young person's mind will turn to love - brother start worrying!

Frank Dailey's Meadowbrook isn't on the Newark Pomptom Turnpike anymore, he now has a place in Newark.

After hearing the latest reports from New York we wonder who's defending this country beside us. Times Square is blacker than ever, but servicemen are a dime a dozen. Let's hope SHE doesn't leave HER heart at the Stage Door Canteen.

The N.Y. Rangers are due to remain in last place for the remainder of the hockey season, and Madison Square Garden is removing the rink for three rings - and we don't mean Ballantine's.

The Phantom is still in the comic strips and Smilin' Jack is still a bachelor. The 8th. Ave. subway still has its crowds and the wolves are still riding the rails. Attention all draft boards; and from latest reports there are still Fifth Ave. busses - remember those balmy June evenings? Dig me Jaxon?

The Polo Grounds is being put into condition and "Sugar Hill" is still on the beam. By the way the "Count" is now in California.

The "Little Flower" has the Air Raid Wardens on the ball, and Guy Lombardo is now at the Roosevelt Hotel.

Glenn Island is closed for the duration, and Jimmy Dorsey with Helen O'Connell (now, now boys) are appearing nightly at the Hotel Pennsylvania and are picking them in. Who said there aren't any civilians left?

Sammy Kaye is at the Strand Theatre, and the good old Paramount has Xavier Cugat and his Latin rhythms, and as an added feature, "Hitler's Children."

Remember when you used to "cut" classes to see T.D. or B.G.?

Washington Heights, the Grand Concourse, and Flatbush (if you want it) have that hungry look, and the gals in these parts are now knittin' for their particular kittin' instead of for Britain.

Penn Station and Track 27 is still there and say is Fort Dix still packing them in! The baloney is really selling these days!

In case you're interested, Fort Worth has four trips daily to New York.

We may all have a "Touch of Texas in Our Walk," but take us all back to New York.

We must be nuts, all this good publicity for nothing,.....well the zoo in Central Park is still "for free"

Keep 'em laced high Jaxon.

P.G.

March 28, 1943 TANK TALK Page 9

"WHENEVER HE GETS MEAN AND INDEPENDENT, SERGEANT, JUST LEAVE HIM ALONE AND YOU'LL SEE HOW QUICKLY HE GETS OVER THOSE CUTE LITTLE SPELLS!"

KEEP 'EM ROLLIN'

(Parody sung to the tune of, "Battle Hymn of The Republic")

I

"Mount Up" comes the order.
"Turn Them Over" comes the shout.
We slam 'em into second
And prepare to move them out.
It's a cinch we'll never live
To die of rhumatiz or gout;
Our tanks keep rollin' on.

REFRAIN

Glory, Glory, Keep 'em rollin'
" " " " "
" " " " "
By God, they've made a tanker out ofme.

II

Our eyes are on the "tach" and trail.
Our feet are on the floor.
Our ears are nearly deafened
By the mighty motor's roar.
Our heads are bumped and battered
And our bottoms sure are sore;
Those tanks keep rollin' on.

REFRAIN

III

We're the roughest, toughest outfit
In this God forsaken land.
We're on a steady diet of
Tumble weeds and sand.
Though our bodies may be broken
Yet our spirit still is grand!
Our "yanks" keep rollin' on.

REFRAIN

R.P.L.

747th. PARTICIPATES IN DISMOUNTED REVIEW

On the evening of March 22nd., the battalion marched in a dismounted review for Major General Bruce, TDC commander. Reliable sources inform us that the appearance of the battalion was very good, and that most of the lines held well,past the stand."

INSPECTOR— INSTRUCTOR APPROVES TENACITY OF BATTALION

Major Ralph W. Sleator in an interview last Saturday on Commando Course number three, stated that the 747th. Tank Battalion had done, "especially well considering the amount of training they have had." "This is not," the Major stated, "faint praise." (He has seen many organizations go over the course).

"Your appearance the first day was ....well......." and he shook his head east and west, "but since then," he smiled, "you've come through."

* * *

MEDICAL DETACHMENT OPERATES WITH SKILL AND DISPATCH

(continued from page one)

the field, saw Pvt. "Angie" Melluso take six stiches in a torn fingure; he certainly can take it. Didn't even grimace.

More vital statistics were these: eight cases of measles (7 in A Co.) Pvt. Anthony Ferraro (A Co.) fracture of the left knee, off the 15 foot wall. Pvt. Albert Bernard (Hq. Co.) hip injury off the 15 foot wall. P.F.C. "Eddie" Ling (C Co.) fractured ankle sustained in Close Combat fighting. Sgt. F. Southern, broken fingure, also on Judo stuff.

These men weren't as lucky as Joe Rapp of Service Co. who fell off the rope platform (25 feet high) lit on his back (he's a 260 pounder) and didn't even chip a fingernail.

MUSICIANS WANTED

All men with musical ability are asked to join the Drum and Bugle Corps (or future dance band now being formed). Those interested in either should see Pvt. Pat.Grimes of Service Company.

COMMANDO COURSE COVERED
(continued from page three)

least prepared to fire. Lt. George Myer, course commander told us the story of one excited recruit (from another outfit) in commando town, who held a pistol, a grenade, and a box of matches in his hands. Exploding dynamite, crackling rifle fire and a steady buzz of machine gun bullets overhead so unnerved him that he lit the grenade, threw it into the street, threw his matches into the building under attack, and then ran completely around the building and fired his pistol point blank at his own men coming up from the rear.

Needless to say, all courses were interesting and officers and men alike found them highly instructive. (Oh yes, all officers went thru the entire ordeal, one in particular with blisters on his feet that would have incapacitated a less determined man).

The proof of the prodding was the lung disrupting 5 mile run participated in by the entire outfit on our last morning at the course. One company made it in 5[illegible] minutes. (Who's this guy Cunningham?)

We might sum it all up by saying the experience left this battalion harder, faster and a hell of a lot more tired than they've ever been before.

When do we do it again?

R.P.L.

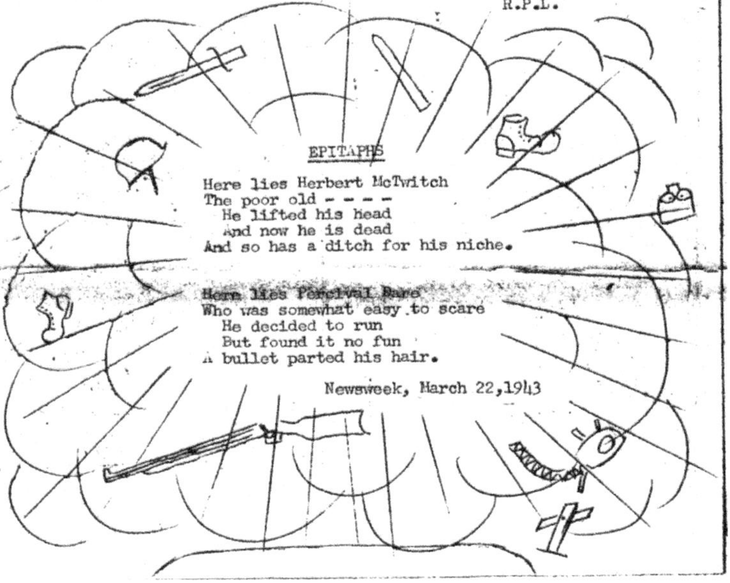

March 28, 1943 TANK TALK Page 12

1st. Lt.: Are you happy in the Army?
Pvt. Draftie: Yes, Sir
1st. Lt: What were you before you joined the Army?
Pvt. Draftie: Much happier, Sir.

- - -

Capt: "The man who sneaked out of the barracks last night and met a girl in the woods will step forward. COMPANY HALT!!

- - -

Lt.: Soldier, I found a pint of gin and a bottle of ginger ale in your locker! What am I going to make of that?
Soldier: A couple of high balls, Sir!"

- - -

Soldiers Voice from Rear Seat Of Taxi: "I say driver, what's the idea of stopping?"
Driver: "I thought that I heard somebody tell me to."
Rear Seat: "Drive on; she wasn't talking to you."

- - -

Why don't you salute me?" barked the irate Lieutenant.

"Gosh, Sir," drawled the bewildered selectee, "they told me never to raise a hand to an officer."

- - -

Officer (in charge of rifle range) 'Hey, don't you know any better than to point an empty rifle at me?"

Recruit Smith: "But it isn't empty Sir! It's loaded."

- - -

Do you remember the soldier who, when asked what he'd done with his wages, answered, "Part went for liquor, part for women, and the rest I spent foolishly."

- - -

1st. Soldier: "I just brought a skunk into the barracks."
2nd. Soldier: "Where you gonna keep him?"
1st. Soldier: "Gonna tie him under the bed."
2nd. Soldier: What about the smell?"
1st. Soldier: "He'll just have to get used to it like I did."

- - -

"Pardon me miss," said the sentry, "but it's against regulations to swim in this lake."

"Well, for heaven's sake!" exclaimed the maiden, "why didn't you tell me before I undressed?"

"It ain't against regulations to undress, lady."

- - -

OH, YEAH?

The boys were griping about the sons of politicians when a new voice spoke up behind them.

"I'm a politician's son," said the voice, "and I'm in the Army just like you ordinary guys."

The gang turned around to see who owned the voice.

"Yes, Captain," they replied.

- - -

Disipline:

To his Negro infantry company in camp, a colored first sergeant spoke dire words: "From now on when ah blows dis yere whistle, ah wants to see a huge, inpenetrable cloud of dust come boilin' outa them tents. An' when 'at dust clears away, ah wants to find three rows of statues."

- - -

See you all again next issue. If you have any of your own jokes, send them in to us for publication.

At Ease Troop—Smile If You've Got One!!!

By Richard Terboss

HUMOR SURFACING DURING MILITARY SERVICE is always welcome as long as it happens in a way that does not disgrace or belittle someone or cause harm. My personal experience with humor in the military is rather limited. I describe my working environment as "working in a bubble." My Army MOS was 96DAO; Image Interpreter. I worked nine hours a day weekly in the states and twelve hours a day, seven days a week in Vietnam. My desk was either a light table for scanning reels of film or a regular table for piecing together conventional photos. The staffs were usually three or four people. We worked in bunkers, a building, or in an expandable van. Add to that I was newly married, living off base when in the states and I was 26-29 years old and working the whole time with others who were 19-20 years old. As you can surmise, there was no time nor interest for hi-jinks. Nevertheless, some light moments come to mind, so buckle up:

KEEPING SOUTH JERSEY SAFE

AT SOME POINT IN MY BASIC TRAINING at Fort Dix (1966) my squad, for no apparent reason, was told we would be pulling night guard duty in individual two-hour shifts. When the time came for the honor, it was pitch dark and it was raining monsoon style. I was driven to my post which was somewhere (or nowhere) in New Jersey. I had a rifle, in retrospect probably blank rounds, a poncho, helmet, a small light and not much else. Did I mention the rain and the darkness? I was told that the structure, taller than a human, in front of me, was an ammo bunker and that it was a walking post. There was no illumination. Did I mention it was pitch dark? I started checking out the structure and my wrist watch.

The shape of the structure was that of a Quonset hut which blended into a raised slope behind it. There was no door, no window, no roof hatch. Funny, I had never seen a Quonset hut with grass cover. Camouflage, perhaps? Did I mention it was pitch dark and raining like a torrent? After two hours my relief came and I went back to the barracks where I dried out. It was clear that this was a test to see what one could tolerate. It was also clear that I had spent part of the night guarding A DIRT PILE! At least I could say "Relax South Jersey, we've got your back."

WINE, PROPERLY AGED

IN 1966, WHILE LIVING IN DUNDALK, MD, a suburb of Baltimore, Maryland, my wife and I lived in a tiny efficiency apartment that was the second floor of a residence. While I was in training at adjacent Fort Holabird to become an MI Image Interpreter, my poor wife, Dorothy was stuck in the apartment all day. That is until our landlords, a nice, elderly Italian couple took a liking to us. The Fazzoli's lived on the first floor. Papa "F" was probably a whole five foot, four inches tall and one hundred twenty-five pounds.

Mama "F" probably reached five foot and was lucky to break 100 pounds. Mama "F" was a great cook and took pity on the young couple upstairs living on 200 dollars a month, including housing and meal allowances. I was still an E2, not even a PFC. Mama "F" would send up a meal now and then and spent time with Dorothy. Papa "F," among other things, made his own wine, both red and white. Here is where the story gets interesting, but first some background.

We lived only a stones throw from the rear gate to the Fort. When walking the neighborhood, one could not escape the strong order of yeast, or what we thought was yeast. Later, we found out that Seagrams operated a distillery in Dundalk, not far from the Fort. They distilled and bottled whiskey and rye there.

One evening, Papa "F" offered us a taste of his home brew red. One normal sip and my temples started to pound. I don't think Dorothy was doing too well either. Upon inquiry, we were told Papa "F" aged his wine in discarded WHISKEY BARRELS that he picked up at the distillery. Although Mama and Papa seemed to have no problem with their product, we were very careful with our sips after that.

SPARROW PIE

IN 1966, WHEN I WAS IN SURVEILLANCE TRAINING at Fort Holabird, our downstairs landlady (the Mrs.) would send up to myself and my wife a casserole supper she called Sparrow Pie. She said that the "Mr." went to Sparrows Point to get his meat. We did not know what that meant and we did not care . The meal was delicious and was free. We did joke about it though.

Sparrow Point, named after the original land owner, was several miles away. It jutted into the Chesapeake Bay, south of the of Baltimore, Maryland harbor. Why would the "Mr." go out of his way to get his meat? Sparrow Pie? Sparrow meat? Nah! Can't be! Unfortunately, we never looked deeper into the question. Sparrow Point, my current research has revealed, was once the location of a huge Steel Mill. In its day, it was the largest in the nation. This was the case when we lived near there but the company went bankrupt years ago.

As I further researched this tale, I checked out Sparrows as a source of nutrition. I found out that Sparrows, especially the species called *house sparrow* can produce almost one ounce of edible meat, each. Furthermore, sparrow meat is considered a delicacy in Japan, and in England and is available in the US. The taste is said to be gamey, like venison. Sparrows are seen as aggressive and a nuisance and are found in great numbers. I can see where they could have been a neighborhood target to the people living near the mill. It could have been a great "snack" location for Sparrows. But Sparrow meat in Sparrow pie? Nah! It can't be!

Still the "Mrs." did call it Sparrow pie, not Sparrow Point pie. I will never know the answer to our dinner content. I'm OK with that.

BY THE BOOK: ALMOST

WHEN ONE IS IN US ARMY BASIC TRAINING, you try to stay below the radar every day, every way. That certainly was the case for this naive country boy from the foothills of the Endless Mountains of Northern Pennsylvania. One of the dictates handed (shouted) out in training was that you salute an officer if one is approaching. Failing to do so would upset the military gods and bring down the wrath of every past and present "circle brim" i.e. Drill Sergeant. Trainees were given one week-end pass during training and I was flying back to my hometown in Hallstead, Pa. I was in travel greens. As I was walking through the Newark, NJ International Airport terminal I saw a uniformed gentleman coming my way. He had some shoulder dressings and looked to me like he was the real deal. As I readied my right arm to offer a snappy salute I realized what he was. I had almost saluted an AIRLINE PILOT. To quote Jerry Seinfeld; "Not that there is anything wrong with that."

TAKING WALKS IN THE SOUTH

IN 1966 MY WIFE, DOROTHY, AND I lived on St Helena Avenue in Dundalk, Maryland while I was stationed at Fort Holabird. The East end of the street ran into Dundalk Ave, the main street in the town. After my school ended for the day and after dinner, we liked to take a walk, heading for an ice cream shop across Dundalk Avenue. In the summer, we would get some ice cream. In the winter, we would get a hot chocolate. Even if there was only an inch, or more, of snow on the road we had to stand back at the traffic light on the corner. Dundalk, in 1966, did not have the means to clear roads. There was no annual safety inspection for car tires either. We "northeners" watched and laughed as cars approached the light and skidded halfway, or more, through the intersection. When the coast was clear we crossed the street. Luckily, we did not witness any accident. Drivers were very careful but sometimes not careful enough.

In contrast, while living in the outskirts of Fayetteville, NC and I was stationed at Ft Bragg, we had a different experience than that above. Our half-house apartment was on Dairy Drive. After supper, and after cleaning up, we relaxed while awaiting the "all clear" sign that meant it was okay to take our evening walk. Dairy Drive had widely spaced houses on one side and pasture land on the other. Usually there were a few horses grazing in the pasture. There was no sidewalk. The street was hard packed red clay. We were waiting for our nightly 7:00 p.m. thunderstorm to arrive. We laughed that you could set your watch by it. Sure enough, it came and it rained torrentially for twenty minutes, then passed. That meant that it was "all clear" to take a walk. The water just rolled off of the clay and the street was smooth and dry. We were very thankful for the drier and cooler air that was left behind.

This Is My First Rodeo

By John Steele

IN THE LATE NINETIES, I was in the Army reserves, and I was doing my first annual training with my new unit. There were six or seven of us on this increment. We were combat engineer instructors, and our mission was to go down to Fort Leonard Wood, MO and work with active duty instructors teaching new recruits who were going through Advanced Individual Training (AIT).

I had not been down to Fort Leonard Wood since I graduated from basic and AIT myself thirteen years earlier. When I was there in 1985, we were not allowed to leave our barracks area, let alone leaving the base and exploring all that central Missouri has to offer. What does central Missouri have to offer besides strip clubs, dive bars, hillbillies, scary tattoo parlors, and alcoholics in the most rundown trailer parks in the country? I was determined to find out. I won't keep you in suspense. Nothing. There is nothing else in central Missouri. It is called "Misery" for a reason.

But even though there was nothing, I still had many "firsts" during that annual training at Fort Leonard Wood.

Most of us drove there. We needed vehicles to get to the ranges we'd be working on, and it allowed us some freedom when we weren't working. It was the first time I drove to Missouri, and the first time I saw and crossed the Mississippi River in my car.

We got there on a Sunday, got settled in, and me and a couple of guys who had been there previous years went exploring. There are numerous caves in the area around Fort Leonard Wood. We drove to this one cave that one of the guys knew of. He had a flashlight, and we walked a good quarter mile inside this huge winding cave. We noticed that there were all these bats that were hanging around the top of the cave. He then thought it would be hilarious to turn off the flashlight, and see if we could find our way out. It wasn't that hilarious. To add to the fun, he threw a rock up toward the ceiling, and got the bats to start swarming around. Still not all that hilarious in my opinion. I didn't start freaking out, but the other guy with me without the flashlight did, so the flashlight was turned back on and we left the cave. It wasn't that late in the day, there was plenty of sunlight, but it was so incredibly dark inside that cave. I did the old "put your hand in front of your face" to see if you can see it. Nothing.

After we left the cave, we got a bite to eat, and the other two guys decided they wanted to get a beer or two. I said sure. That was until we pulled into the parking lot of Big Louie's. How to describe Big Louie's? Picture a traveling carnival, or a big gaudy used car dealership, with three big search lights circling up across the sky, flag streamers draped across everything, lots of balloons, loud music, and tasteless billboards. It is lit up so much, you can see it for miles. It is the size of a small grocery store, complete with a cheap, tacky gift shop with all the sex toys and porn you could ever want, and a strip club. It does not have a bar, as I was told they lost their license to serve alcohol. But what a coincidence. Right next door, in the same parking lot, is a liquor store. Patrons can leave the strip club, buy a bottle, drink it in their car, and return to the strip club. So convenient.

I talked them out of Big Louie's. Too big. Too loud. Too busy. I told them I wanted a quiet place for just one beer and relax before our first day instructing recruits. They other two guys I was out with were both married. I was single. They called me a pussy, and we went elsewhere.

Next we hit what looked like a nice restaurant/bar. It was another strip club, but much cleaner and classier than Big Louie's. To be fair, Big Louie's set a real low bar. I'll call this new place a gentleman's club, it sounds better. I've never been a fan of these types of adult establishments, and I protested. They pleaded for one beer, so I reluctantly agreed. As the other two busily put dollar bills in G-strings, I shyly sat there, sipping my beer, trying not to blush, trying not to stare. Some scantily clad tropical looking girl came up and sat next to me. She was the cutest, most modest looking girl in there. She introduced herself as Lalani, and she was Hawaiian. She asked why I looked so quiet and timid. I stammered something, trying to look her in the eyes, and not look at anything else. She asked me if I wanted a lap dance. I politely declined. But my buddy said I did. "No, I'm good."

"No, he wants one," and he pulled out some bills and paid Lalani.

I was given an awkward and unemphatic lap dance. I think she was as bashful as I was. She later told me that it was her first night ever working as a dancer, and she wasn't sure if it was for her. But she heard the money was good, and this place wasn't skanky like most. She sat and we chatted for a bit, but then she told me that the boss was giving her the stink eye, and she better get up and circulate some. A few minutes later, she came by and told me that if someone were to buy her a drink, she was allowed to sit with them and drink it.

Obliviously, I mumbled, "Oh..." and just sat there. After a moment, she mingled away, and my buddy kicked me. "Dude! She's saying if you buy her a drink, she is allowed to sit with you. Buy her a drink." I was in my early thirties, and totally clueless. I bought her a way overpriced drink, she milked it for well over an hour, and we sat and chatted.

Finally, after a couple of hours, my buddies were finally ready to go. Lalani gave me a big hug, and told me she hoped I would come back and see her.

I did. Well, I tried. On a couple of nights we went back to that club. She wasn't there. Each night, after a beer, they wanted to go somewhere else where the beer was cheaper. We hit some very seedy bars. Some had strippers, sleazy, trashy, crude. We hit this one raunchy place, I still remember it was called Daisy's. It was so sordid, the girls didn't even bother dancing. They just flashed you or sat with their legs open and urged you to give them money and accompany them in some back room.

I was done. I told the guys I was leaving. And since I drove that night, unless they wanted to walk back to the base, they were leaving too. I wasn't doing any more bars or clubs or back ally dives. Nothing. Finished.

Believe it or not, I still haven't touched on the most memorable moment of that annual training. We worked Monday through Friday, and had the middle weekend off. What were we to do? I wasn't doing any bars again. One of the active duty instructors we were working with informed us of this one small town that was having a rodeo on Saturday. I had never been to a rodeo. None of us from my unit had. So the whole bunch of us headed off to this small town rodeo that was about an hour away.

It was really hot that day. The beer flowed freely through just about every person in attendance. And for a small town, damn near every single person from that town and every neighboring town must have been there. I had never seen so many pick-up trucks in one place in my life. They had rows and rows of porta-potties, but the later in the day it got, and the more beer that was consumed, the less people bothered to walk to the porta-potties. They went behind bushes and behind fences and behind pick-ups. It was probably safer. The later it got, and the more beer that was drunk, the more some of those porta-potties mysteriously got pushed over. Some people even stopped bothering to go behind something to pee.

I wasn't overly impressed with all the horse riding and cow roping things that they do in a rodeo. The steer wrestling and the bronco and bull riding was...livelier. Like a train wreck. You don't want to watch, but you can't look away. It was an experience. A whole new world. Hillbilly heaven.

It was getting close to the end of the evening, and the final event was called, "The Circle of Death." The later the day got, and the more beer that was consumed, and the more porta-potties got pushed over, the more buzz there was about this circle of death. Let me explain. There are these circles about ten feet in diameter, placed all over the floor of the rodeo arena. Everyday people, drunk cowboy wannabes, each pay money into a pot to stand in one of these circles. Then some crazed bull is released into the arena. The last person to leave their individual circle wins the pot of money. How freaking crazy is that? All of the contestants of course had to sign a waiver releasing the rodeo from any liability.

There were a dozen or more circles in the ring. They would announce a guy's name, and say what *podunk* town in Missouri they come from, and the guy would trot into one of the circles acting as tough as can be. Then they announced the last person, Clarence Donaldson from New York. Sergeant Donaldson, from my unit, had entered the circle of death.

Let me tell you about Sergeant Donaldson. He was African-American. One of the only African-Americans seen on the grounds that day. Definitely the only African-American standing in the circle of death. He was in his late twenties, and about as urban as can be. He was from a poor, inner-city neighborhood. Where every other person standing in one of the circles was white, with a denim shirt, a cowboy hat and boots on, acting all cool and tough, Donaldson is out there with a New York Knicks Jersey on, sneakers, and a baseball cap on backwards, shaking his butt and dancing to the music. The announcer was making a big deal about him being from New York. People are booing like crazy. Donaldson has his arms in the air, egging them on. At one point, he pretended to moon the crowd.

Then the bull was released from the chute. He charged out and rammed the very first cowpoke he came to. OH...MY...GOD... This poor white guy must have flown back fifteen or twenty feet, landed in a thud, and didn't move. I wasn't even sure if he was breathing. It was like a Mack truck hitting a mannequin. The bull then turned directly towards the next nearest guy, and hesitated just a second. The guy didn't move. It was like his feet were stuck in cement. He was frozen. The bull flipped him up a good ten feet into the air like a rag doll. The guy did a one and a half somersault and face planted into the dirt. The bull turned towards a third guy. This guy ran. Ran like his life depended on it, because it did. This guy vaulted a five foot fence, and was gone. About half the other guys were gone too. The guy who face planted was up and gone. The first guy who got smashed was at least moving a little bit now, but he wasn't getting up. Some rodeo official was on horseback standing between the bull and the guy who was writhing on the ground in pain. Meanwhile, the bull was milling around the fence where the one guy escaped. The bull looked pissed that he missed him.

There were now only half as many fools standing in their circles, so many of them ran. And what comes out but a whole bunch of rodeo clowns. They've got bright clothing and clown make-up on, and they are bouncing around acting all goofy, trying to get the bull's attention. It took what seemed like forever, maybe five or ten seconds, for the bull to be coaxed back towards where these ignoramus rednecks and Sergeant Donaldson were standing. And who are they trying to get the bull to target? The crazy, black dude from New York. The bull charged at Donaldson, and somehow Donaldson juked and dodged out of the way, while managing to stay in his circle. The bull proceeded on and took out the Missourian who was behind Donaldson. These dumb white guys didn't seem to try to get out of the way. It was like they were rooted in fear. The latest victim was lying crumpled and bloody. The bull turned around to get a second run at the poor guy, but a couple of the rodeo clowns intercepted the bull, and steered it away while a couple other clowns dragged the latest victim to safety.

Time to get the bull re-engaged. Once again, they tried to get the bull to take aim at Sergeant Donaldson. It was like they decided there was no way they were going to let this black guy from New York walk away with the money. During the next passage, Donaldson wasn't so lucky. The bull's horns were rounded down some so as to not be quite as pointy and lethal, but it still nicked Donaldson on his backside and belt, lifting Donaldson off the ground. It didn't fling Donaldson. Worse. Donaldson's jeans were now caught on the horn of the bull, and Donaldson was now literally flying by the seat of his pants. The bull ran for a few steps, then lowered his head and started raking the ground. Using Donaldson as a plow, the bull tilled a small trench in the dirt of the arena. The rodeo clowns intervened, and with his pants half ripped off, Donaldson exited the arena.

Donaldson was okay. He was sore. His face and chest were a mess, cut up and scrapped like road rash from the bull trying to use him as a hoe. He said he had no idea the bull would be so big. He said the bulls looked a lot smaller from up in the grandstand. We asked him why he did it. He didn't do it for the thrill. He thought it would be easy to win and he'd make a little extra money. He's fast and athletic, and he thought that most of these Missouri boys looked fat and slow. He was perplexed on how he didn't win, and felt it was unfair that the rodeo clowns put a bull's eye on him. But wow, what a memorable first rodeo.

Many of the guys from my unit have this tradition where they go to this upscale German restaurant at least once during their AT. I had never eaten at a German restaurant, so I was a little skeptical and worried. All I could picture was weird sausages and sauerkraut. Then they started telling me about schnitzel. The name alone doesn't sound all that good. Then I was told it is deep fried pork. Pork is already greasy. You're then going to deep fry it? I was nervous, but I tried it. I loved it. Another first, eating German food.

While we were getting ready to leave this German restaurant, who do I run into? Lalani. She was out with one of her girlfriends. We chatted for a bit. I told her I went back looking for her. She had decided that dancing was not for her, she only worked the one night. She asked if I wanted to have dinner with her the following night, and I did. We went to this Mexican restaurant, and talked for a couple of hours. I found out her real name was Darlene. She used the name Lalani because it sounded so much more exotic. It was nice meeting up with her, but I was leaving the following day, nothing was ever going to come of it. But I still count it as the one and only time I ever dated a stripper.

So let's recap all of my firsts. It was my first annual training with a new unit. It was the first time I drove to Missouri. The first time I got to experience Fort Leonard Wood as a free man. I visited my first cave that wasn't turned into a tourist spot. I had my first experience with bats. I had my first lap dance. I visited Big Louie's and Daisy's and a few other really scuzzy dives. It was the first and only time I've been propositioned by a prostitute. It was the first and only time I've been to a rodeo, though it was the first of many times I've seen porta-potties get tipped over. It was the first time I've seen the circle of death, and I hope I never see one again. I saw guys get pummeled and flipped and used as a plow by a bull. I saw my first rodeo clowns. I learned that I love German food, and I very briefly dated a stripper. What a fun filled two weeks, made possible courtesy of the United States Army.

Lighter Side of Offloading a C-130 at Lunch Time

By Gary Redlinski

1969 Army Mortuary, Vietnam

IT WAS LATE MORNING when we got a call that a C-130 was about to land with about twelve remains on board. Bob and I headed to the flight line to meet the plane coming in. The plane had just arrived when we got there. It was a bit unusual to see one of the Air Force fire trucks arriving as we drove up. Bob and I wondered what was wrong. As they lowered the back ramp of the plane, there was a very bad, putrid smell coming from the plane. When we backed into the ramp and got out of our *deuce & ½* truck, the odor got stronger. As we started unloading the stretchers from the plane, the airman in the ramp was trying to cover his mouth and nose as he was breathing. Most of those body bags had really decomposed remains. Some of those bags had leaked and green fluid had leaked off the pallet under them.

Bob and I just started transferring the remains to our truck. We could see why the fire truck was there, to hose out the belly of the plane. That poor airman just stood there and watched us unload. He looked a bit pale, with a funny shade of gray-green.

Bob and I "being the nice guys" we were, started talking, loud enough for our airman to hear, about going to lunch when we got back. The more we talked about being hungry and what we might get, that airman started gagging a bit.

We finished loading our truck and wished the airman a "good day." He couldn't wait for us to leave. We started to get back in our truck and the guy was already waving for the fire truck to move in and do their hosing out.

When we got back to the mortuary and unloaded the truck, we did go to lunch.

transfer truck

The Fly By

By Arash Rivazfar

(Aboard the aircraft carrier USS Ronald Reagan, CVN 76)

I REMEMBER THIS TIME when we were in the Mediterranean Sea. I was a 'yellow shirt.' (A Sailor must earn the aircraft director qualification. Those personnel work in the ship's air department and specifically handle and maneuver aircraft.)

During the height of night flight ops (operations), one of the pilots decided it would be funny to buzz the tower in an EA-6B Prowler. I was half asleep at the time because I had been directing all day with only three hours of sleep. It scared me so bad I pissed my pants a little, and when we had our nightly meeting, the Air Boss noticed a darker color on the front of my pants said, "I think you had an accident, Petty Officer Rivazfar."

I laughed and said it was the pilot's fault in the Prowler that buzzed the tower. And I need sleep!!

Little Stories

By John Michael (Mike) Ryan

These stories describe some of what I learned in boot camp and beyond (1969-1973). We complained, even joked, about these "learning" experiences. Today I just laugh about them.

THE COIN-FREE LAUNDROMAT

At boot camp at the Naval Station at Great Lakes, Illinois we were responsible for keeping our underwear clean and dry. I learned how to tie a regulation knot in order to attach my skivvies and undershirts, via string, to my personal, short clothesline which was about two foot long. This outfitted line was then laid across two parallel clotheslines running outside and alongside the building. They were just out of reach height wise. They were about two feet apart. Clothespins would have been something else to keep track of, so they were not used. We had "wash" squads. I was a member of one of these because I was unable to reach the outdoor lines. The "clothesline" squads were made up of taller recruits. Our Commander felt this routine was the best use of manpower and the best way to maintain both good hygiene and schedule. My clothes were marked with my serial number (B168289). Identifying mine when a clothesline squad member took clothes down was easy for me.

QUICK-STEP MARCHING OR LEFT, RIGHT/LEFT COMBO, COMPLETE THE RIGHT, REPEAT

I learned how to half-step or quick-step when marching in order to save my heels from injury. I suspected I was in trouble when I was selected to carry the battalion flag (old 539), when in formation. Turns out my predecessor did the job poorly. I guess I was qualified because I was more nimble. I did not know why the flag bearer had to be short and it was not my first choice. It did not take long for me to recognize the problem. We had to maintain a thirty-inch stride. That meant the six-footer behind me was always stepping on my heels. To save my heels, I learned to quick-step by quickly moving my left foot a few inches forward. A mini-second later the right foot came forward and kicked the left foot the rest of its way with the right foot then continuing its journey. I became so good at quick-stepping that no one noticed, or at least no one laughed. Except for that six-footer behind me who kept trying to step into me, no doubt on purpose.

MAKING DO WITH GOVERNMENT ISSUED CLOTHING

I discovered this dilemma early on. What do you do when you are five foot, two inches tall and weigh one hundred ten pounds? This is only two inches and ten pounds over the Navy minimum. My shoe problem I solved by putting cardboard in them. If you recall the fictional character Captain America, the loose fit of my other clothes made me look like him when he was at his own fictional boot camp. This was before he took the fictional magic serum that made him bulk up into a super hero. I had no magic serum but I was able to get some tailoring done. Nevertheless, my fatigues looked like there was room for another half-person plus myself. When I got assigned to a ship, I was able to buy tailored dress whites and blues from someone who was leaving the service.

GUARDING A DUMPSTER

From dusk to dawn, we in our barracks at boot camp had to pull two hour shifts guarding a dumpster that was alongside the building. We had to stand at parade rest with a rifle and no ammo of any kind. We never figured out why. What was the threat? We were told of coyotes, rats, or a drunken sailor trying to escape. Go figure. But we were superb in our execution.

Second Amendment Adventures—A Tale of Four Chapters

Preface: The Event That Didn't Happen

ONE FINE MEETING DAY for the Rochester Veterans Writing Group a bunch of good veterans were minding their own business, talking about what would happen if somehow, sometime, someone made the area form a local militia actually mentioned in the Second Amendment to our U.S. Constitution. The conversation was light but serious, and based on the writing prompt we had been working on that Saturday meeting-day morning.

What a fabulous, far-out idea. The right to bear arms! The humor about what would fall out from forming a militia! What 10,000 good and bad turns would happen if someone tried to form a regional militia! That idea was much too great to let go by. I had to write that baby up, likely to the astonishment of my comrades.

IMPORTANT DISCLAIMER! The above paragraphs are the only nonfiction showing up in this four-part series, *Second Amendment Adventures*. The events in these four following parts did *not* happen, but if they did, this tale might well follow this guide:

Chapter 1: The Only Good Part: It Wasn't Raining

Chapter 2: Henry Vth Addresses the Troops

Chapter 3: David Niven Forms an Army

Chapter 4: The Light at the End of the Gun Barrel

The names have been changed because, well, they aren't anyone, anyway. How much safer can that be? Fun, though! Look for Chapters 1-4 scattered through this book. Now Chapter 1:

Second Amendment Adventures-1

The only good part: It wasn't raining.

Chapter 1

By Vaughn Stelzenmuller

SEVEN O'CLOCK ONE TUESDAY MORNING, and the phone rings.

"Good morning Mr. [redacted]," said the shaky voice from the other end, "this is Richard Johnson…from the City Council, but you can call me "Jake," with an unmistakable, nervous choke to his voice, "and we, *uh*, the City Council would ask if you could help foaming, *uh,* ***forming*** the militia group we need to set up quickly. You see, we,…

[Oh, I get it, it's that dumb-ass odd new State program, which "mandates localities set up militia programs to serve various local and state defensive needs." Whatever blazes that means. What now?]

"… we would like to invite you because of, *um*, your past honorable military service, and need an experienced leader to set up this program," he coughed, "and we have a contingent, *uh*, group, *uh*, of *citizens* who will need direction. Can we ask you if you will help?" Johnson finally asks.

This certainly falls into the ugly-surprise range of how a morning would start. And certainly, citizen-me would feel obliged to "help out." But why the heck would I want to jump into this wasp nest without finding out what the &@#$@#$!! this militia crap involves? I doubt ever.

"Certainly, I would be glad to *help out*," I somehow reply firmly, the Charles Bronson in me rising to the surface, "so when would any sort of planning group be starting?"

Sigh. *I should have suspected that this was starker, darker than some kind of bureaucratic "planning group" caper. First, for a City Council member, a 7:00 a.m. call would be like an oh-dark-four a.m. panicked call for a civilian person. Second, clearly the city council* president *didn't want to make a scary, risk-laden phone call like this, so they "deputized" council member Richard "Jake" Johnson to bite their bullet.* Some *junior member had to be the Council's bullet-bait.*

"Well, Sir," Johnson continued, a whisp of breath-catching, "there actually is a militia group...forming in our new downtown open space, and we need to...to...get started."

"Ah, I know the area well. So you need my help right away," I keep pretending Charles-Bronson confident, "and want me to assist with the organizing? When should I start? And please, don't call me 'Sir.' You're in charge, so I should call *you* 'Sir.'"

This is all certainly a weird beginning. What could go worse now?

"*Um, umma,* well, this new militia is forming...right now...downtown," my new superior stumbled, "so is there, maybe sometime this morning, we can take you downtown? We actually, *uh,* need you...to *take charge.*"

I think I just heard this right. The city, or somebody, *has invited a huge neighborhood of, god-knows-what/who to show up downtown to form a militia gathering? And which means that some previous sucker for the job just got fired. What could go worse now? What! Serves me right for turning off the local news all the time.* To continue the saga:

ME: I can be ready in 15. Pick me up in front of the house, and just tell me what I should bring with me? *[That sounded pretty confident and* ex-military *didn't it? Doesn't mean much here, but might keep me from being fired before even getting hired for, god-knows-what]*

RICHARD "JAKE" JOHNSON: ....Great! *Uh,* we can pick you up in fifteen minutes, *uh, fifteen,* as you say! Oh, you don't have to bring anything. *[Did I just detect utter surprise and relief in the voice coming over the phone?]*

ME: I'll be ready.

[all right, I know very well that a government limo with council member couldn't possibly make it before 45 minutes. Gives me time to pack notebook & pen, comb my lack of hair, brush teeth, and put on suitably intimidating armed attire for, god-knows-what lurking in the details around the corner. I hope they gassed up the car.]

Councilman Johnson is overjoyed to see me ready for transport downtown, and on the trip to god- (oh, you know!). I review whatever details I could about what situation I would deal with in about ten minutes. He knew I had left the Service as a captain, and that meant (to him) that I could work with a large person gathering, somehow. He didn't know much else, so I figure this was a seat-of-the-pants (i.e., governmental) setup.

Then we arrive, and Councilman Johnson and I step from the car.

Whoof!

Batting neither eye, my mind sweeps the scene at the newly cleared park, and thought I had been time-traveled to between the World Wars. The scene looked like those black & white film footages of those forlorn veteran soldiers camped near Washington, D.C., the so-called Bonus Marchers, waiting for General MacArthur to ride in with troops on horseback to drive the veterans off. My rough estimate of person-power in that park tells me we were heading for some kind of catastrophe.

Once out of the car, I make sure to hold onto my civilian superior's arm to keep him from dropping back behind me, and then calmly observe, "I see roughly eight- to nine hundred people here. My last rank expected me to command maybe 250 people, well-trained of course, and the group in front of me would require a lieutenant colonel level of experience, assuming they were well trained," I begin, not looking at my companion, but scanning the horizon-crowd slowly, and as we say in the military, *in a military manner*, "so we will definitely benefit from your staying by my side to help make this, *uh*, project work.

"I do hope I'm wrong when I say I see a huge number of our—citizens—already armed? It would be good to know what fraction of those firearms are loaded." *Dyspepsia City*, ahoy.

I could swear that my civilian superior, three-eights my age, almost collapsed seconds ago from a combination of near-death dread, fear of how badly he might throw up, and fervent wishing that he'd never run for office. And I truly need the lad's presence to give this situation force. For myself, I would need a combination of Robert Redford and David Niven to pull off my half of all this. And, I need to keep Johnson upright.

The best part about all this was that it was *not* raining. The rest of the day could easily keep tumbling downhill. Quickly.

[TO BE CONTINUED: CHAPTER 2, *Henry Vth Addresses the Troops*, p.130]

FIRST THINGS FIRST: FAMILY!

Always Gonna Be a Country Girl

By Sue Spitulnik

I WAS RAISED IN THE RURAL FARMING AREA of Western New York State. It was the 1950's. We learned to walk by holding on to the dog and making mud pies in the garden was a real pastime. We played outside from the time we could run and even used party phone landlines. I was comfortable being a country girl. When I got into my teens I didn't take the time to learn how to put on make-up so it looked even, dealing with the fact my lips and eyes aren't even. I learned that no one's are, but most ladies have more patience for such things and wear some make-up, or a lot.

I left my hometown six months after graduating from high school to become a low-maintenance military wife, and soon a mother. Who had time for make-up and fancy hairdos? I guess one can make time for anything they want to do, but being a fusser was never on my radar.

My Air Force husband was an enlisted man. Often called a grunt by any officer. It wasn't a putdown, it was reality. As an enlisted man's wife, I only spent time with other enlisted men's families. It's the way of the service. No mixing of the officers who were the bosses and the enlisted. It's still that way.

This was fifty years ago, and maybe my memory isn't quite correct, but I remember the officers' wives as generally being high maintenance, or at least higher than I was. They wore make-up, nicer clothes, and to me, were snooty when mixing with us low-lifes while doing things on base. One day in the bank, a lady strutted in and walked to the front of the long line at the teller window. A discussion ensued and she revealed she was a Major's wife and didn't have to wait in line. Thankfully someone explained to her, that his rank was not hers, and she went to the back of the line. A wife is just a wife.

When my son was born in 1973, back in the days a new mom and baby had to stay in the hospital for 72 hours, there was some loud yelling coming from one of the rooms. Being the young and nosey age everyone was, we moved to our doorways to investigate the commotion. One of the doctors was telling a woman that she was indeed going to take care of her baby. She yelled back she was a Captain's wife and she was not going to care for that boy, because she had wanted a girl. A few more words were exchanged, and then some loud sobs were heard. We all watched the doctor storm down the hall, and in a few minutes the mom, who was in full make-up, with her hair done, shuffled to the nursery, in her expensive-looking robe, to retrieve her son to take back to her room to feed. Now that might have happened whether the female was an officer's wife or not, but she tried to play that card. Not one of the other mothers could grasp her attitude. We couldn't wait to cuddle our babies.

Being a young mother of my wonderful son and then adding a daughter was my pleasure and focus. If I wanted or needed to get something done away from the house, I took a shower, shook out my short hair, and was ready for the day. I even heard my husband's friends comment they never had to wait for me when we decided on the spur of the moment to go someplace.

Looking back I realize even after getting divorced I never dated an officer. Who knows why. Maybe my small-town ways, lack of adorned physical appearance, and being interested in the job I was doing rather than how I looked while doing it, didn't attract any officer's eye.

I'm now married to a great guy who was in the Army and Army Reserves way back when. He was not an officer either. He grew up in the same rural area I did. We are both more interested in the other one's heart than in appearance and are much richer for it.

Though the puddle of my life's experiences continues to expand, almost into an ocean, no matter where I live, at heart I'm always gonna be a country girl.

One of Those Days

By Vaughn Stelzenmuller

IMAGINE A MEETING OF EX-SOLDIERS who gathered to write experiences based around *prompts* from one of the members. It had been General [redacted]'s turn to ask, and the prompting had to do with, "What did we do with our *down* time."

To clarify, I asked her, "All right, General, you want to know if we sometimes filled our spare un-military time with stuff like reading, shooting the (or spending money on) shit, going to movies, or trying to sleep? Well, yes."

"'Yes', *what?*"

"Yes, Ma'am."

* * * * *

"You see, General," I continued, "there was this one especially rotten non-combat, but completely-military 1971 day when I had got chewed out by three bosses in a row up the chain for a stupid non-combat decision involving military housing. It (and *I*) really were stupid, and the deed also served to shred up what was left of an otherwise sunny non-combat afternoon.

"Before heading home I decided to stop by a small grouping of fellow junior officers to run the bad events by these good semi-friends, excuses and all, knowing they would be sympathetic, Ma'am. They weren't. They all laughed a bit though.

"Recovering the pieces of the day by finishing up a good library book at home seemed like a splendid idea. A phone call home, though, informed me that I would instead be escorting my full-military wife to the post movie that night (or else), though there might be a few minutes for that book later. *Not* helping to assuage my grief at my *increasing* grief—THE #@#%@!!!! post movie was *Lion in Winter*, a long and tedious tale involving King Henry II and Kathryn Hepburn at each other's 12th Century throats. For therapy I decided to snag an order of German-style French fries on the way home. I was not overweight, but I could have used the few bucks spent more than the *blankity* greasy things anyway.

"King Peter-O'Toole Henry didn't get deposed, his three sons wanted to more or less kill him, his oldest one named Anthony Hopkins was in love with the king of France, and Kathryn Eleanor of Aquitaine Hepburn got sent back to her captive castle for Christmas. With that much movie joy, exhaustion naturally crept in, so I didn't feel like picking up that book later, and it was off to sleep. Except I couldn't. Damn day.

"Sometime around *oh*-two-hundred dark o'clock my zero-and-one-half year old daughter started crying. It was 'my turn.' This little babe often had a problem keeping food down during her first year, and I suppose she wanted to remind me of the fact by coating most of the crib sheet with her earlier dinner. Hell, I'd cry my buns off too to get out of a ghetto like that!

"With the stamina and incentive that comes with the age-20s father grouping, I was able to first pick up my daughter and cradle her in my left arm and clean her face. That went well, and then I found myself not only wiping up the crib right-handed, but also pulling off the old sheet, fitting a new non-fitted sheet onto the crib padding, squaring away the pillow and two stuffed animals, all the while holding my completely relaxed sleeping package. No wrinkles in the sheet. This was a *military household*, of course. Somehow, with all this fiddling around, my daughter had drifted off from the moment I lifted her from her self-made cotton sewer, past the moment I put her back down onto a five-star hotel heaven, and then *tippie-toe'ing* from her room. It's good to be the hero.

"You see now, General, my answer to your questions was *YES*. One of those all-the-above days, and you also know why it was *hands-down* the best one of my young military life."

The Screaming Summer of 2003

By Lee Nelson

I WAS BROKEN.
We were both
broken
but could she
be broken?

She had learned
to scream
with exceptional
success
elsewhere.
Co-parenting.
It happens.
It's challenging
and Linda Blair
arrived
for the summer
and I was hardly
a man of God.
I was hardly
a man at all.

She would scream
screams
perhaps heard
on D Day
or in
factory accidents
or in the distant wings
of asylums
screams

from the 9th circle
of Hell's fluffy
pink bedroom
with a chronically
strangled
happy
bear
in a purple dress
screams
that would leave
Buddha
dragging a smoke
on a shoreline
with a 5th
skipping stones
defeated
incredulous
to reason
and she
would scream
some more.

Allowing her
to scream
without
acknowledgment
wasn't working.
She was winning
as I would
stand
at the
ends
of my property
to hear
what the neighbors

could hear.
The cops arrived.
They were
incredulous
too.

I knew few parents
other than my own
and the parenting
advice
of the
non-parent
is as
useless
as the
battle advice
of the
non-combatant
and I received
lots
of that
advice
and she
would scream.

I recruited neighbor
parents.
I knocked their doors
like a man knocks
a car window
with a cup.
They offered
about as
sparingly
with the same

averted eyes
and all the same
indifferent
empathy.

My young son
began
to grow old
with me.
One day
finally
in a perfectly
dirty
white tank top
he entered
her pink
fluffy Hell
drew his arm
back
toward himself
in an untaught
backhanded
fashion
and clocked
her exhausted
blushing cheek.

She stopped.
She stopped
ever so
briefly
and we rejoiced
in the brief
moans
and hiccup

coughs
and guttural
exhausted
bewilderment
of perhaps
a goat
with an
upset
stomach
and in all my
brief
gusto
of suppressed
joy
I yanked his
guilty arm
and him
into the
hallway
and was at a
complete loss
for words
but preferred
silence
anyway
when he said

"Do you hear that, Dad?
Do you hear that?"

"Yes! Yes, I hear that,"
I replied.

"What are we gonna do, Dad?
What are we gonna do,"

he asked
demanded
begged
beyond anyone's
years
of undeserved
suffering
of me
of an indifferent world
of an indifferent god
or an absent
child
psychologist.

"I don't know, son.
I don't know,
but we
can't
do
that,"
I replied.

I called him son
so formal
so desperate
so never before
or since.

I sent him away
on his bike
hoping he'd
return
and grabbed a beer
and turned on
the tv

and began
to cook
Sloppy Joe
as she chose
to scream
again

until finally
like a faucet
she just
shut
off
and presented
hungry
in the kitchen
with her
strangled
delirious
smiling
purpled dressed
bear
wiped her snot
and tears
and said

"That smells good,
Daddy."

I drained the burger fat
into the sink
with hot water
and I was
drooling.
I don't think
she noticed

but I'm pretty sure
the bear did

and we both sat
as I made
plates
of
Sloppy Joe
and
Kraft
mac n cheese
and peas
and my son
arrived
just in time
hungry
sweaty
ready
to eat.

There were smiles
on all of us
as we ate.
We chewed
and looked at
each other
with the most
genuine smiles
ever to
fruition
within a family.

These battles
won
are the greatest

triumphs
of love
but they're
vile
victories.

We don't talk
about them.
We don't
know
how to

and this
explains
so very much
family
dysfunction.

This explains
the incredulous
as odds are
none of us
are incredulous.

This explains
parents
that smile
and drink beer
and have advice
even good advice
even the exact advice
you need
but no
capability
to articulate

any of it
and no place
in the soul
to consider
room for it.

Knowing
what you didn't
do
is too close
to what you
almost did
and that is all.

Walk the wire
and cope alone.
Good luck
my friend.
Good luck.

This stuff explains
the best laughs we have
before we die.

See Ya Later

By Charles F. Willard

ON VALENTINE'S DAY, 1964, my wife Lynn and I moved from a rented walk-up apartment in Mrs. Messervy's antebellum "single house" in downtown Charleston, South Carolina, to the first home that was truly ours. It was a brick-clad three-bedroom colonial ranch, built especially for us in "Oak Ridge Estates" in Charleston Heights. They had brought the rudiments of it on a truck on the 26th of December, and it was ready in less than two months. Just in time, too, because Lynn was eight months pregnant with our firstborn.

The names "Oak Ridge" and "Charleston Heights" gave a false sense of altitude. Anyone familiar with the South Carolina lowlands around Charleston knows that most of the land there is only a few feet above sea level. Oak Ridge Estates, elevation 29 feet, was a new development of starter homes being hacked out of a swampy forest near Charleston Air Force Base, a short drive for me when it came time to fly a mission.

It was a pretty area. Tall live oak trees draped in Spanish moss gave shade to the houses and some privacy from the traffic on Dorchester Road close behind the new homes. With careful grading, the developers relegated the swamp to the ditches along the sides of Oak Ridge Drive. And we hadn't experienced any flooding, even with the torrential rains common in that coastal area.

Ours was the first home built in the new neighborhood, but soon there were others. Before long, an Air Force captain, his wife Maryann, and their two boys, Allen and Sam, moved in just down the block. Jim was a C-124 pilot and was gone much of the time on overseas missions. Maryann taught school and typically arrived home about a half-hour after her two kids, leaving the boys on their own for that short time, a pretty typical situation in those days.

The boys were ten and twelve, anxious to explore their new domain. It didn't take them long to discover the ditch. While mowing my new sparse lawn with my new hand-pushed reel mower, I watched as they headed down there after school with their fishing poles. After a few minutes, they flung the poles down and ran back to their house as fast as they could, emerging in a moment with their dad's fish net.

Off to the ditch they went again and flailed around in there with the net, trying to capture something. I couldn't imagine the ditch could harbor a fish big enough to be worth their attention, but something sure had fascinated them. A bullfrog, maybe? After a lot of splashing, they had whatever it was netted and headed again for the house on the run. I was too far away to see what they had, and they disappeared into the house and remained there for ten or twelve minutes.

Soon, out the door they came again, headed back down the road with the net, apparently hoping for more of the same. As the boys disappeared down the road, Maryann arrived home from the other direction, parked in her carport, and went directly into the house. I don't think she noticed her boys around the curve by the ditch.

Barely five minutes elapsed before Maryann, barefooted and wrapped in a robe, came flying out her front door, screaming at the top of her lungs. I ran toward her, wondering what awful thing she had discovered.

"Ch-Chuck, th-there's some sort of m-monster in my bathtub!"

The boys, hearing her scream, had come running, too.

"Allen!" she yelled, addressing the older boy, "What do you know about this?"

"Aw, gee, Mom, he's just a baby."

"Get him out of there! Now! I'm not going back in there until he's gone. And you clean the tub, too!"

"Aw, Mom...."

The boys and I marched into the house, ready to confront the "monster." Allen and Sam had put a few inches of water in the bathtub, and there, his large mouth agape showing lots of needle-like teeth, his short little legs clawing at the side of the tub in a futile escape effort.... was a baby alligator.

I took the net from Allen and, after a few tries, re-netted him and started for the door.

"See ya later, alligator," said Allen.

"After a while, crocodile," said Sam. There was distress in their voices, but it was hard to tell if it was because they were sad about losing their pet or fearful of what would happen next.

I wasn't privy to that conversation. I was walking the netted alligator far, far down the road, to a place where the swamp was still undisturbed.

I Lost My Son

Sue Spitulnik

WHEN IN NATURE I have a pretty good sense of direction. I can get back to where I started whether going for a walk, looking for leeks, or when I was younger, hunting. There's nothing like the peacefulness of the woods to sit, listen, and reflect.

On the other hand. Driving to unfamiliar places could be an adventure, or a hassle depending on how you look at it. When my son was six or seven he would get angry when I got lost while behind the wheel and refuse to get out of the car when we got to where we were going.

In 1986 I was living in Tacoma, Washington, and had taken my maiden name back after getting divorced from my active-duty Air Force husband. Expo '86 was happening in Vancouver, British Columbia, Canada. A world's fair was something you attended if at all possible. Hotels were booked solid almost a year in advance so private citizens who had extra rooms with a private bath were renting them out. Pullout couches, cots, and mattresses on the floor became home for a couple of days for a lot of people. My mother type friend, age-wise, Leta Neros, and I loaded up my kids and headed north on I-5. Jeremy would have been thirteen, and Molly, eleven. When it came time to cross the border into Canada we were asked to pull into a waiting area. We were delayed while the authorities called my ex at work, on a military base mind you, to verify he knew I was taking the children out of the country. If our last names had been the same it probably wouldn't have raised any red flags.

In those days a driver used a map, landmarks, and written instructions to find a new destination. Oh, and a pay phone if necessary. We found our lodgings the first time with no trouble so we could meet the house owners at the expected time. After that, when trying to return to the house from the fairgrounds, I seemed to turn the wrong direction out of any parking lot and head south instead of north. First, we had to realize I was going the incorrect direction, and then we had to get turned around to go the right way. Getting back to the private residence was a challenge. It was frustrating after a long tiring day, and I wasn't used to getting lost that easily.

At Expo, because it was a world's fair, many countries had displays in individual tents. With your admission ticket, you received a "passport" booklet with each country having its own page to be stamped when you visited. It became a thing with the school-age children to collect as many stamps as possible. Jeremy figured if he went to as many exits of tents and begged for a stamp, he would be the one to collect the most stamps. He was thirteen. I gave him permission to take off on his own while Leta, Molly, and I returned to the purple zone sitting area where we had enjoyed lunch to wait for him. He was to come back there when he was finished.

I don't know whose idea it was to section off the fairgrounds into color zones by painting walkways, but it was a good one because each food court looked the same. There were at least six zones, maybe more. The problem arose when we ladies returned to our lunch spot and saw that I had told Jeremy the wrong color. I think we changed zones after he left us. I was thirty-three, still a country mouse, and scared out of my wits I would have to go to security and explain how I had lost my son. I could imagine the hullabaloo when they called my ex and told him, or even worse if they made me call him. Thankfully it was way before kids were being snatched for trafficking. I don't think that thought ever entered my head.

I don't remember how long he was lost to us, but I do remember feeling hopeless and fearful. I was fortunate to have Leta, my level-headed friend with me. She kept telling me to give him some time to collect his stamps and that he was smart enough to retrace his steps back to the landmarks he remembered and ignore that the walking area colors were not what Mom had said.

After he had gotten at least ten stamps, he retraced his steps, expecting to find us in the purple zone. That's when he felt a pang of anxiety because we weren't there. He told me he then started looking for familiar tents and recognized a ride he and his sister had gone on. He went to the fellow working there and told him the situation. The worker was my angel because he listened to Jeremy, knew the grounds well, and brought him right to us. I was never so happy to see my son.

I sure was glad no one had to call my ex that day and tell him the circumstances. I don't know if the kids ever told him or not. We laugh about it now.

U.S. Army Story

By Richard DeLisle

IN JANUARY 1959 I JOINED THE U.S. ARMY for a three-year tour of duty. I did my 8 week basic training at Ft. Dix, NJ and then went to the Corp of Engineers at Ft. Belvoir, VA for my 12 week training in special electronics repair. I specialized in night vision equipment and mine detectors. It turned out that the instructors were the top students from the previous class. I was number one, so I got to teach the next class. I spent most of my time repairing equipment that was returned from all over the world.

My discharge from the Army was set for January 1962. The practice had been to release you early if your discharge was around a holiday, so I expected to get out just before Christmas 1961 and start college in January 1962.

In May of 1961 I was told that I had to choose between two "hazardous duty" assignments overseas that would last for 5-6 months and then I would be released early when I got back. The choices were both with the Corp of Engineers in a support function. One was in a jungle in a place called Viet Nam and the other was on the ice cap in Greenland close to the North Pole. I chose Greenland.

In early June 1961 I took a complete physical and was bussed to Ft. Dix for a flight to Thule, Greenland. Our non-stop trip to Thule didn't happen. On the way, we lost one engine and then another of the four. We were forced to land at a base in Goose Bay Labrador and stay there for a few days until they got the plane fixed.

Finally, we took off and landed at Thule Air Force Base in Greenland. We stayed at a base near Thule called Tuto. While there, we had more health checkups including a complete dental exam. We were also issued all the clothing that we would need for the stay.

It was 138 miles from Tuto to Camp Century all across the ice cap. The soldiers leading us were in vehicles called polecats that looked like oversized, enclosed snowmobiles. We were in enclosed trailers called wanigans that had sled runners. The wanigans were pulled by diesel tractors. Travelling at a few miles per hour it took forever to get there. The way was marked with flag poles so as to avoid deep crevasses. Not long after we left Tuto, we ran into an Eskimo family and we stopped to talk with them. I took some pictures.

When we arrived, we saw a series of above-ground long tent-like structures and we were puzzled because we were told that we would be living under the ice cap. It turned out that the tents were built to house the crew that built the under-ice camp the year before. The tents were to serve as emergency housing should something go wrong under ice. They were stocked with cots, blankets, stoves and fuel, food, etc.

We were told that our mission was to support the civilian research teams that were conducting experiments to see if people could live on the ice cap if they had to. One major project was to construct an under-ice railroad from Century to Tuto so that supplies and personnel could be transported easier. More than 50 years later my daughter Jackie did some research and found articles that identified the real purpose of Camp Century. It was being built to house a series of under-ice ballistic missiles that could be randomly moved to alternate spots on the rails. The rails would not go to Tuto. Google Project Iceworm to see much more on this clandestine effort.

I was a specialist 5th class (sergeant) with a crew of electricians assigned to me. We were assigned work by the base commander and by a civilian research team out of Vicksburg, MS called Waterways Experiment Systems. In reality, this turned out to be a U.S. Army team.

One project was to lay and power lights along the runway so planes could land during the winter when it was 24 hours of darkness. They were going to attempt to stay at the camp through the winter. This was crucial because they did not like to leave the nuclear power plant unattended for so long. We had trouble keeping the lighting system up and running because the arctic fox kept eating the wires!

We also had to light the top of the weather tower so planes didn't hit it during the darkness.

At the under-ice camp, there was a main trench with tunnels all along the sides. In the tunnels, were prefab houses where we lived. They were very comfortable. All other needs were in separate buildings that you walked to. We worked 12 hours on and 12 hours off. When we got there, it was 24 hours of daylight and when we left, it was 24 hours of darkness.

Some things that I remember from the assignment:

• When I took pictures, the shutter would freeze open from the extreme cold.

• I read letters from home for a soldier who could not read.

• I wore a dosimeter to measure radiation from the nuclear reactor but we were never told the results.

• I got major frostbite many times.

• We had to run the diesels 24/7 because, if you shut them down, the oil and fuel would freeze and you could not get them started again.

• Planes would come and go with personnel and supplies; they landed on skis; sometimes they froze to the landing strip and we had to go out and push the plane sideways to break it loose from the runway.

• On august 24, 1961, a Catholic chaplain flew up to conduct church services for us. We had church services every Sunday that we conducted with our own volunteers or with "the real deal." After the service, six people left in the helicopter that they flew up in. Another plane on its way up with supplies, noticed debris on the ice cap. It turned out that the helicopter had crashed less than a mile from our camp. All six people aboard, including the chaplain, were killed. We sent a team out to recover the bodies and we kept them in the main trench for a week until the debris could be examined and the bodies taken back to Thule. I have a copy of the article about the crash.

Because of the cold temperature in the trenches, we were able to keep food in the trenches rather than in refrigerators. Milk was frozen and put on the tables to thaw. You could have all you wanted to eat and everybody, including the CO, worked in the kitchen when needed.

• Rank was not enforced and all were treated as equal.

• There was a major problem with the ice shifting in the tunnels so when we got there, many men were assigned with chainsaws and sleds to carve back the walls that would push over the buildings if not shaved back. This was very labor intensive.

• The reactor had many operational problems and was not supplying power for much of the time that we were there. Instead, three 300kw generators were used for power. The reactor also gave off much more radiation than expected so we all had to wear dosimeters to measure the daily radiation exposure. We were never told how it was going. More shielding had to be added to the reactor.

• We had three different types of boots depending on what we were doing that day and where the work was.

• Buffalo radio station WKBW held a contest to see who could send in a picture from the farthest away of a box they supplied with their call letters on it. I took a picture and was going to send it in but when I showed it to a friend, he noticed that my wedding ring was showing! Elaine and I had eloped and no one back home knew about it yet. Close call. I didn't send the picture in but I still have it.

Some key items from a U.S. Army report:

• A temporary camp was built to house the personnel who would construct the under-ice camp. The temp camp was built between 14 June and 10 July 1959. The temp camp went on to serve as an emergency shelter for the under-ice personnel should something go wrong.

• The construction of Camp Century was started on 12 May 1960 and was completed in October 1960.

• The waste from the reactor was disposed of in many ways: liquid waste was pumped deep into the ice cap. Solid waste was sent to the coast, entombed in concrete, and dumped in the ocean. The reactor was disassembled in 1964 and sent back to the states. During the disassembly, soldiers were exposed to much higher radiation than expected so they had to shorten the work days and use more men.

• Liquid general waste was pumped under the ice cap. Solid general waste was buried in a trench as in a land fill.

• The camp was abandoned in 1966 due to the inability to control the shifting of the tunnels.

• A film crew re-visited the camp in June of 1969 and took pictures of the damage that shifting had caused in the last three years.

When I got back to Fort Belvoir in late October 1961 I expected to be released soon. Instead, the Berlin crisis started and all personnel were held in place with no releases being granted. I wasn't released until March of 1962. I had expected to start college in January but had to delay it until September.

I have some pictures that I took in Greenland and an envelope from Camp Century that I sent to Elaine when she was practice teaching in Fillmore, NY.

Our marriage, while on duty, was an experience in itself. I had always heard that if you wanted to marry a girl it was important that your ask for her parents approval. I asked her father and he said no!

In New York, you had to be 21 to marry. I was 20 and Elaine was 19. We decided to elope to North Carolina where the marriage age was 18.

I went to the local Walmart and bought a set of wedding rings for $200. I put $20 down and paid $20 per month. Smallest diamond ever!

On the first weekend of October 1960 Elaine flew to Washington from college. I borrowed a friend's car for the weekend and we drove to Halifax, North Carolina. All that was required for marriage was a mini-physical; because I was in the Army, I didn't even need that check.

We were married by a justice of the peace. We drove back to Washington and Elaine flew back to college. Our families didn't know we were married.

When I came home from Greenland in October 1961, Elaine was doing her practice teaching in the Southern Tier of New York State. I drove there to pick her up and we went to my parents' house. We told my parents that we were married, and they were very happy for us. We did not tell her parents because of her father's feelings.

Her parents were going to surprise her by picking her up to go home for the weekend. When they got there, they were told that she had already gone home with me. When they got home, they called my house to see where she was. My mother answered and told her the good news. Needless to say, they were very upset.

Her parents insisted on a religious service, so her pastor married us in his house during November 1961. We were married twice with no divorce! We have always used the October 7,1960 date for our anniversary. On October 7, 2023, we celebrated our 63rd anniversary.

The Bomber Suit

By Charles F. Willard

IT WAS THE POST-WORLD WAR II 1940s, and at least in our small town, radio was still king of the entertainment world. Flush with pay from his new traveling salesman's job selling animal health products for the Borden Company, Dad brought home first, a new 1947 Buick Super—with a radio(!), then a radio-phonograph console for our living room, and shortly after that, one of the newly available "clock radios" for Mom and Dad's bedroom. The expenditures seemed so profligate to my mother that she feared we might have to reinstitute our wartime diet of meatless Tuesdays and Spam the other six days of the week.

Dad wasn't worried. He knew he could dream up a new idea to bring in some extra money. But Mom was apprehensive. Experience had taught her that, more often than not, Dad's schemes involved extra work for her. Like the time during the war when Dad, who was too old for the draft, decided to go into the chicken and egg business with his younger brother, Bill, who was 4-F because of his eyes.

In addition to laying eggs, chickens also generate lots of chicken manure. It has to be shoveled. It turned out that Uncle Bill, a city boy, "didn't do chicken poop." Who got stuck with that little chore while Dad was away all week traveling and selling? Yep, my mother, the farm girl.

Back to the radio. Combining a clock with a radio was an innovative idea that had just become commercially available. Instead of waking to a jangling alarm, you could shake off Morpheus to the soothing sound of music, or in Dad's case, Deacon Doubleday's farm show on WSYR, the flagship station in Syracuse, where he could hear farm news, complete with a detailed rundown of the latest hog prices on the Chicago Stock Exchange. Dad was a faithful Deacon Doubleday listener until Henry Boston, our neighbor down on the John Peet Road, turned him on to Kay and Lem Lemmon's cornball farm program up the dial on WNDR. Up to then, WNDR, a weak rival to WSYR, was held in low repute in our family due to the sleazy quality of their programming—WNDR carried Giants games. Dad, a Brooklyn boy at birth, was a Dodger fan.

But Henry was insistent. Kay and Lem were "funny." They told humorous stories about the herd of goats they kept on their farm outside of Jamesville, a Syracuse suburb. In fact, Henry said, they were so persuasive in praising the benefits of keeping goats that he had just purchased two from them. He mimicked the Lemmons' sales spiel: "Think how much you can save not having the milkman come to your door every morning." And "Goats will eat anything, so feeding them is no problem. Goats' milk is naturally homogenized. It's good for people with ulcers. You can make a bundle selling it to those people. They'll beat a path to your door."

And Dad fell for it. Forsaking the Deacon, he twisted the dial to Kay and Lem on WNDR and was convinced. Who was the salesman here, anyway?

In a few days, I found myself riding shotgun in the new Buick on a trip over to Jamesville and the Lemmon farm. As (bad) luck would have it, we arrived at milking time. It was the perfect situation for an amplified "in the goatflesh" sales pitch. Surrounded by all the goat farmer accouterments (feeding stations, milk buckets, a barrel full of grain, a scale to weigh the milk, a big chart on the wall to keep track of each animal's production, and an extensive pasture with a five-foot fence), Lem and Kay were ecstatic and effusive about their goats, lauding the animals' personalities, ease of care and their profit-making potential. And when the goats produced progeny, you could either raise them as milkers or sell them for meat. "They taste a lot like lamb, but milder and without all the fat."

I could see the dollar signs in Dad's eyes. His new money-making scheme had sprung to life. He never thought to ask, "If goats are so great, how come you're selling so many of them?"

Have you ever seen the back seat of a Buick after two goats have spent a cavorting hour enjoying the ride? It never again smelled the same back there.

By the time we arrived home, I had sort of made friends with the two nannies, in spite of the fact that they had both tried to eat my hair. Dad stopped the car opposite the door to the back porch, eager to show Mom the beginnings of his latest business venture. Mom came to the door, wiping her hands on her apron, trying to suppress the "Oh, no!" look on her face. She hated to discourage Dad. He worked so hard at his extra money brainstorms. But they usually seemed to make more work for Mom and often ended up as "extra expense" extravaganzas.

Dad grabbed one goat by the collar and directed me to take charge of the other. Dad started for the barn with his charge, and mine started me for Mom's flower garden at a full gallop. Half a dozen mouthfuls of peonies disappeared before Mom was able to intercede. The two of us managed to drag the resistant animal into the barn, its cloven hooves plowing up little gouts of backyard grass all the way.

We were in the goat business.

Mom, one of five children, grew to maturity on her father's farm in the days before plumbing, central heat, and other modern conveniences. Milk came from "Old Bossie," the family cow, and all five children learned how to milk her. So, it was Mom who taught Dad how to milk the goats. Dad assumed it would be Mom who would do goat care and the milking while he was away all week. That was how it went for a while, then Mom became pregnant with my middle sister (it turned out—no ultrasounds in those days), and that put the *kibosh* on her goat chores. She was too busy changing diapers.

At nearly 12 years old, there I was, naively waiting in the wings. Dad taught me how to milk. It was fun directing a squirt or two into the barn kitty's face, watching her joyously let the spray go into her wide-open mouth and down her throat, then licking the excess from her whiskers with obvious enthusiasm, meowing for more when the initial stream of ecstasy subsided.

Soon, I learned there was more to the goat business than a squirt. It became my first real job. Occasionally gathering eggs when we were in the chicken business didn't count. This was heavier-duty stuff. Goats had to be fed and milked twice a day—before breakfast and school, then after supper. When winter came, water had to be carried, and the chores had to be completed, below zero temperatures or not. I would struggle through the dark winter mornings and evenings, plowing to the barn through one or two feet of snow, my frozen feet in oversized galoshes, my shivering body clad in layers of holey, hand-me-down sweaters, and jackets. I looked like a many-colored Pillsbury dough boy. They saw me coming, poor goats, and cringed. "Here comes old icy fingers!" How would you like your most tender body parts to be suddenly in the grasp of ten freezing digits?

I protested my lot bitterly, but there was no help for it. I was stuck. That clock radio had sealed my doom.

Then one cold winter day, Dad brought home the bomber suit. It was the greatest gift I received throughout my childhood, and it changed my life because it changed the way I *thought* about myself. Clad in that suit, I became a different person. I could *imagine*! Let me explain.

In those days, our heroes were the crews of the Eighth Air Force, many of whom fought and died high over Europe. Beset by cannon fire from swarms of German ME-109 and Focke Wulf 190 fighters and shredded by intense, lethal bursts of flak, they flew true on their courses, dropping their bombloads on Third Reich targets. They soared at altitudes up to 30,000 feet in unheated, unpressurized B-17 Flying Fortresses and B-24 Liberator bombers and the temperatures inside those aircraft could be as low as 50º below zero. The crews were protected from freezing by highly insulated "bomber suits." Here's a picture of a B-17 crew in their regalia.

Fleece-lined leather helmets, jackets, leggings, and Mae Wests. Not shown are the fleece-lined flying boots.

Dad had brought all the pieces home except the Mae Wests. Somehow, somewhere he had met a fellow who had been a B-17 tail gunner in the war, had struck up a conversation, as he always did (being a salesman), and conversations at that time often got around to the war and what one did in it. The fellow was a little guy, barely five feet tall, not much bigger than me. Tail gunners had to be small. Not much room in the cramped tail gunner compartment of a B-17. When Dad learned that the man still had his bomber suit, a lightbulb went off. He could keep me warm while I milked. Oh! There were fleece-lined mittens too. Now the goats could be greeted by warm hands. I think Dad paid fifty dollars for the whole shebang. This goat business was getting expensive.

Thinking back, I'm amazed to realize what that gift did for me. First, it let me know my efforts were appreciated. But most of all, it let me see my twelve-year-old self differently, not as the downtrodden goat milker, but when I donned that suit, as the World War II hero. *Here's the tail gunner, shooting down Messerschmitts and Focke Wulfs. Here's the B-17 pilot, guiding his craft through fields of flak to his target.* I suppose wearing that suit cemented my dream of someday becoming an Air Force pilot—even if I did look like a chocolate Michelin Man as I slogged out to milk the goats.

The Disappearing Buddha

By Richard Terboss

DOROTHY AND I MOVED INTO A SMALL APARTMENT in 1966 near Fort Holabird, MD. As newlyweds we had no decorations or much of anything else. Everything we owned just barely fit into our small car. My sister had given us a beautiful gold colored statue of Buddha. She meant for it to be a gift that would bring happiness, protection, love. He was about twelve inches tall and sitting cross-legged with a reflective stare. We laughed because it was the only item we had on display and he certainly was not a military man.

Somehow, someway during our many moves, Buddha turned up missing. Many years later, my sister gave us a twelve inch tall concrete statue of Buddha or of one of his many reincarnations. To this day, he resides in the flower garden outside the front of the house. The original Buddha never returned. I would like to think that he is somewhere else being "one with the world." We felt that he should be credited with seeing us through our service time, unscathed.

11 General Orders for Sea (Service) Story

By Noel Nichols

A STORY MUST:

1. Begin with the phrase, "this is a no shitter" or "this is a true adventure."
2. Must be told in 3 minutes or less.
3. No sleeping on the Listening Watch.
4. Can not use more than one profanity in five consecutive words.
5. No denigration of any service rates you are not qualified for, or persons identified in a story by name.
6. Only stories about Military service time.
7. No recycling of others' stories.
8. Speaker must use proper Service Terminology.
9. Speaker must be given full attention while speaking.
10. Listeners must agree there are no major factual errors.
11. If really a true story, time can be extended till finished.

Copyright 2001 @ and compiled by the no-shitter rules work party USS Sullivan's Field Day TinCan Sailors of the scrapped, sunken and lost fleet, 7/29/01.

The Pastor Who Wore Khakis

By Father William B. Leone

HERE IS BACKGROUND to my stories. I joined the New York Army National Guard (NYARNG) as a Chaplain in May of 1980 at the age of 31 years. I began serving with the rank of First Lieutenant and I retired in 2008 with the rank of Lieutenant Colonel.

I wanted to do more for the young people I met as Pastor of St. Jerome church in East Rochester, NY (1977-1980). I felt that my opportunity to provide counsel, spiritual guidance, forgiveness and absolution would be enhanced if I could catch young people, healthy or wounded or in some other way broken by their military service. I was motivated to do this because I was deferred from military service during the Vietnam War. One of my grammar school friends came back from Vietnam having lost both legs. I felt I should do something by way of service to our country, if now I could, and so I applied and was accepted as a Chaplain.

* * * *

My own father served in the Philippines during WWII. Perhaps his service was also inspiring me to do something by way of service. Freedom is not free and the cost must be paid by us in whatever way we can!

I was activated for service in the US Army in 2006 as a Chaplain attached to the 29th Infantry Division out of Boston and Virginia. Historically, this division was involved in the D-Day Invasion of Europe at Normandy. The mission I would now be involved with as Chaplain concerned a Multi-National Peacekeeping presence in Kosovo, on the Balken Peninsula. You may recall the war that broke out the in the 1990's. That war caused much suffering in the form of ethnic cleansing. The Orthodox Serbs represented 8% of the population; The Muslim Kosovars represented 91% of the population.

I was given a driver (Chaplain Assistant) who was armed. I was not. Sergeant B_K_ and I became fast friends! He was a wonderful young man, a husband and father and very committed to his military service. Since the time we served together he rose to the rank of Sergeant Major. It had taken me five tries to pass the US Army driving test for the right to drive a motor vehicle in Europe. After finally passing the exam, we ramped up to deploy into our theater of operation and I was fortunate enough to be assigned a driver, SGT K_. As so, after passing I never once drove a car in Kosovo. Good thing, because the cause of most deaths while we were there resulted from vehicle accidents, civilian and military. Maybe the LORD was trying to tell me something when I couldn't pass that driving test.

In Kosovo the 29th ID was tasked as a peacekeeping presence for the people. There was a danger that hostilities could break out again. And so the constant presence of soldiers from other countries on a peacekeeping mission helped maintain calm in the streets and villages throughout the country. There was one particular bridge in the northern section of the area of concern that was considered a "flashpoint" if things were to "go south." This required a constant presence of one company of infantry from the brigade to always be there. They were supported by armor and by air support (Blackhawk and Apache helicopters). On one of my visits to this location I was bombarded (excuse the word) by a series of young infantry who wished to go to confession. I thought this was a little unusual for our soldiers to be doing in such large numbers. This was so until I realized my Chaplain Assistant, after speaking with the young soldiers, was advising them they'd better get to confession on the double! I had to counsel SGT K_ that it might be better if he prayed for them rather than order them (he outranked most of them) to confess. A few of them I suspect had not been to confession since their *first* confession. One can't help but think that the Supreme Commander HIMSELF may have smiled.

All Kosovo children wanted to leave Kosovo and go to Italy, or somewhere, for an education. One young man, whom the soldiers were aware of, had saved up 100 Euros. This was a fortune in those days. He was directed by others to go to a certain building and there a gentleman would make arrangements for him to leave. He turned over the money and was told to come back later. Unfortunately, when the lad came back, both his money and the gentleman were gone. American soldiers hearing of his plight, raised money to give him a second chance. The outcome is unknown but what is known is the smile that must have swept over the heart of these infantrymen, who were far away from home, away from loved one and in a country that they had never heard of and in a dangerous spot in the world. Kindness knows no boundaries, if you let it.

WHAT COULD GO WRONG?

What Could Go Wrong? A Trilogy

By Dave Cole

WHEN I FIRST WENT INTO THE MARINE CORPS, I got myself involved in some incidents due to my young age, 19. I was initially drafted but then enlisted in the Marines realizing I would be headed for a nasty war in a place called Vietnam. Things happened due to my immaturity with a certain amount of stupidity thrown in.

During my first day at Parris Island, medical tests were performed and vaccinations were given to protect against various diseases. Additionally, urine samples were taken. I entered the Men's room at a Parris Island recruit induction building to obtain the urine sample. After completing my sample in a plastic cup, I exited the Men's room and joined several other recruits also holding their glass of urine waiting to hand them off to medical personnel. I said aloud, "What are we drinking boys?" A Drill Instructor yelled, "Shut up, you piece of shit." I quickly was identified by this Drill Instructor as a wise-ass and I felt sure I would suffer the consequences in due time.

Later we were escorted to the Quartermaster and received our uniforms and other military equipment. Then we all received haircuts and looked idiotic with completely bald heads. We were marched to a barracks and assigned various bunks and foot lockers for our gear. Throughout the experience, the same Drill Instructor seemed to stare at me with his evil eye look.

After putting our gear away in our foot lockers and being taught how to make our bunks, smokers were allowed to form a circle outside and enjoy one cigarette. We had to face outward in a large circle to smoke and no talking was allowed. After I finished my smoke and placed the butt in the can in the middle of the circle, I ran back inside the barracks and stood in front of my bunk. Suddenly another recruit ran up to my bunk and shoved me away from it stating that it was his bunk.

A fight ensued until the Drill Instructor, who already didn't like me, pulled us apart. Later, I was arrested by the Military Police for fighting and taken to the base brig.

The first night in the Brig, a guard forced me to run in and out of the cold shower naked as a sort of brig initiation. The water was ice cold but I eventually became used to it. I think the guard could tell and ordered me to dry and get dressed.

I spent a few days in the brig and was finally sent to Correctional Custody Platoon for five days, Correctional Custody Platoon, or CCP as it was called, was a place where troublemakers or childish recruits were placed to instill some maturity in them before reassignment back to a regular training platoon.

I guess I was still a target as I felt the staff still treated me harsher than other young men in CCP. For instance, one chore we were involved in was moving a pile of sand as big as a large house from one area of the lot 50 yards further away to a different spot. Every one of the new Marines was allowed to carry one bucket of sand on each trip to its new location except for me. I was forced to carry two buckets each time until the entire pile of sand was moved. The buckets of sand were heavy plus it was very hot. Other tortures we had to endure were several trips through obstacle courses running and climbing various objects to tire us out. I was quickly learning it was probably better to keep my mouth shut during training. We were also instructed to keep our CCP barracks clean by using toothbrushes and towels to clean the decks or floors of our barracks. There were no weekend breaks for CCP occupants and it was a relentless task of keeping us moving and sweating.

After a week in CCP torture, I spent another week in Motivation. A place with similar techniques to get my adolescent mind into a more military condition. The one technique used to instill teamwork utilized a large telephone pole where eight men used the pole for "up and on shoulders" techniques. The pole was heavy and I noted the shorter men suffered the consequences of the pole's weight when lowered to the waist and the taller men suffered the weight when it was raised up and on shoulders.

After my time in the brig, CCP, and Motivation, I was assigned to a new platoon to return to Basic Training. I had now learned to keep my mouth shut, not get into any trouble, and work as a team member, the hard way.

I succeeded in the Marine Corps and rose to the level of a combat Squad Leader in the Vietnam War despite my initial immaturity and stupidity.

****

AFTER A COUPLE OF YEARS OF FLOUNDERING when my Marine duties were complete, I became a cop. I spent forty-one years in law enforcement with the New York State Police and Steuben County Sheriff's Office. There were a lot of serious and even deadly situations I was involved in, but also some funny ones.

I was a young Trooper working radar on Interstate 390 in Wayland, NY, about five miles from my substation. When running radar I typically allowed 10 to 15 mph tolerance in a speed zone because I knew most vehicles typically displayed speeds from 3 to 5 mph inaccurately.

On a particular summer afternoon in July 1979, I observed a vehicle traveling at a speed of 82 mph in a 55 mph zone. I pursued the car which was transporting a family.

I typically pulled a vehicle to the adjacent shoulder with my red lights displayed. I would park my Troop car behind the vehicle I stopped, angling it slightly to the left behind the stopped motorist to ensure my safety from traffic passing by when walking to the vehicle.

As I approached the vehicle, I observed a man in his forties as the operator. There were several other adults in the vehicle. Upon arriving at the operator's vehicle door, he rolled down the window and immediately started a rant of extremely foul language basically asking me why I had stopped him. He continued with foul, disgusting language indicating I was an annoyance. I politely asked for his license and registration and informed him I had stopped him due to his speed of 82 mph in a 55 mph zone. He produced the requested documents but continued with his insulting rhetoric.

A lot of times the issuance of a summons is discretionary. It's the officer's decision pending the initial interview. I can tell you after his profanity and abrasive attitude, this guy was getting a ticket.

After reviewing his actions I advised him to sit tight as I was going to issue him a summons for speeding. This is followed by more disgusting language from him.

Walking back to my patrol vehicle I determined that I would treat this 'gentleman' with a ticket causing him a large fine and a raise in his vehicle insurance. I was fuming but gloating with satisfaction over my decision.

Upon reaching my patrol vehicle, I discovered I had accidentally hit the door lock button when exiting it, I couldn't believe it! All the doors were locked, my red lights were in operation and my vehicle was running. I was so embarrassed. Upon a vehicle stop, it is necessary to advise Dispatch of your location and the reason for the stop. This procedure gives Dispatch a 'heads-up' on your location and situation in case something arises. Dispatch would know there could be a significant problem if they don't hear back from you in short order.

I had no spare key as it was left at my substation when I went on patrol. I was screwed. I had no way to contact Dispatch and they would soon sound the alarm and the Calvary would be coming even if from afar. I do mean the Calvary and without a doubt, I was totally screwed.

I had no options. I had to find a way to get back five miles to my substation to retrieve the spare key. There was little traffic on Interstate 390 and my only option was to request this obnoxious operator who I was about to summons for speeding for a ride back to the State Police barracks in Wayland so I could get the key. It would be difficult to serve him tickets for vehicle and traffic violations and expect a five-mile ride, actually ten. He had to take me there and then return me to my vehicle.

I swallowed my pride and asked for a ride which he agreed to. I quickly retrieved the key and was returned to the Troop Car. As I approached the police vehicle I heard Dispatch state, "All units, 3E25, Trooper Cole's vehicle has been spotted abandoned at milepost marker 29 with no signs of the Trooper. All units respond. Ambulance and area Police have also been notified. Please keep the radio traffic clear due to this emergency. Zone and Troop Commanders have been notified. Lieutenant Mills is *en route* to SP Wayland."

I was in big trouble.

Note: At the time this situation was no laughing matter. But now, remembering it, I can see the humor. It was almost as an immature mistake as shooting off my mouth my first days in the Marines.

And finally, years later I had an experience of redemption and healing.

I spent time in the Marine infantry in Vietnam. I killed many a man in gunfights either in precision gunfire or haphazard firing from fear. My life before joining the State Police was marred with thoughts of a war where I watched friend and foe die either immediately without notice or with pain and suffering.

My mind has always carried memories of war and I've detailed it in many stories of horror. Some say I don't detail my stories with a lot of substance, but then their memories don't have the pleasure of my actual combat experiences. Maybe I shoved my descriptive language aside of how I felt, how the scene appeared, or how men anguished in pain at their final moment.

I'm no author or writer. I'm just me. I describe my experiences as best I can.

I seem to dwell on battles and death.

However, in my grammatical distress let me tell you a time I saved a life. I actually saved a person who surely was meant to die before my eyes.

It was sometime in 1986. I was a State Police Sergeant in Wayland, New York, my hometown. I heard a call over the Wayland ambulance channel alerting medical volunteers that a woman was suffering cardiac distress near Route 63 in the Town of Wayland. I knew that logically any response beyond eight minutes in a cardiac arrest situation was futile. I was within five minutes and trained in some medical assistance.

I responded to this residence and arrived simultaneously with a female Trooper from SP Wayland. The residence was the home of the parents of friends of mine. Upon entering the home, I noticed an older woman lying face down on the kitchen floor. I checked her carotid pulse and found it absent. I checked her breathing and it was non-existent. She was clinically dead. I stripped her shirt and bra off as I turned her on her back. I started CPR by placing my palms a couple of fingers above her xiphoid process and started compressions. I instructed the Trooper assisting to perform the required breaths after my set of compressions.

I had never tried to save a victim from imminent death during the war except in futile attempts to aid a comrade dying from an explosion or gun wounds. Then I desperately tried to stop the oozing blood and assure the soldier everything would be okay.

This elderly woman lay unresponsive as I feverishly compressed her sternum to motivate her heart to restart. The Trooper with me and I worked as a team. It seemed like an eternity passed and I was getting frustrated that my efforts were unsuccessful.

What good was I doing in this profession? This victim wouldn't respond to my efforts. What was I doing wrong? I finally had a chance through divine intervention by my close proximity to this dying or dead housewife and grandmother to many of my friends. I could enlighten my own spirit here years later to save a life instead of taking a life in war.

Desperately, I continued my compressions but was tiring quickly. I was frustrated as I felt I was failing. The ambulance and paramedics still hadn't arrived and I felt I was elected as this victim's coroner instead of her savior. I was getting more and more disheartened.

Suddenly I felt her fingers move as they were next to my inner thigh as I knelt close to her to do the compressions. I felt her breathing start and her body slightly move.

Just as suddenly a paramedic entered the kitchen and told me to step aside. The ambulance arrived and they scooped the lady up and transported her to Noyes Hospital in Dansville, NY.

I had done it with help from the Trooper with me. I had saved a life instead of taking a life. I felt wonderful.

I later received a wonderful short letter from the woman thanking me for saving her life.

One of my greatest moments as a New York State Police. I still have her letter.

Battalion Surgeon Journal Winter 1945

By Joe Mele

You may have heard war stories from relatives, or friends, or told some yourself. Many veterans are reluctant to talk about their military experiences. It can be painful to relive some events.

Infrequently, but memorably, a photo, or a phrase, or the voice of an Army buddy would trigger a memory, and my father would relate a specific WWII incident—no doubt cleaned up a bit. These vignettes would always be preceded by the admonition "War is hell," and many times, a current of dark humor flowed through his story.

The following brief narrative is a fictionalized amalgam of some of these stories, presented to you as a day in the life of a Battalion Surgeon. I hope some of the humor comes through.

******

0600:

It's quiet enough to get a letter off to Tessa and Jeanni:

\- - - - - - - -

Somewhere in France

My dearest darling Tessa + Jeanni,

Yesterday, I received 2 letters from you. They certainly lift my morale. They are just what the doctor ordered (substitute of course). I live just for your letters. Through them I live in my happy home once again.

Sometimes when I help censor other companies letters, I see that they too miss home + wives + family but they are all not as fortunate as I am. I can count on a letter every day, or almost every day. That means a lot to me darling. Keep it up.

Last night, we slept in a hay stack + had some sort of chicken mites all over us + on our bed clothes when we got up. We itched and scratched all over but are gradually getting rid of all our guests. Still, we're the lucky ones.

I'll say it one more time – the front line is hell. The toughest men I've ever encountered have been broken at the front. Yes darling, mail from my wife is the morale booster, the best there is. I can just see Jeanni holding her bottle and me holding Jeanni +giving out with the baby talk + my darling Tessa sitting next to me holding my arm and just beaming at her two babies.

So with that daydream in mind I will say all my love and kisses for you + Jeanni.

Your ever loving + admiring husband + daddy.

0700:

Must have been a lifetime ago that the 749th slogged onto Utah Beach—tanks bogging down, rain washing diesel fumes into the mud. The brass said the beachhead was secured, but it was still FUBAR. By the calendar, it's been eight months since that day at the beach.

There's frost on the ground and the sky's clear, some clouds in the distance. No aircraft buzzing. No shelling. No machine gun or small arms fire. There's a large creek near the barn we bedded down in last night. It's not completely frozen over, so water is still flowing. I kneel on the bank to fill my canteen and I spot a fat corpse (a Jerry by the looks of the helmet) hugging a boulder forty yards upstream. I toss in a couple Halazone tablets, shake it up and take a swig. Ahh, nothing like the taste of chlorine in the morning. Two years ago, I would have retched at the thought of drinking cadaver water.

As a second louie, I went by the book. When my rotation came around to inspect the Mess in training camp, I put on a clean white glove, reached in and swiped the bottom of the biggest pot. If it came up greasy and it usually did, I ordered "washem' again." "Yes sir" was the reply, but you'd think I told them to shoot the family dog. Two promotions later, I got to know the NCO in my medical detachment fairly well—like a younger brother. Sergeant Mack offered this candid advice: "Captain, you're too GI."

Yes, the Army has rules for everything, the written ones, and the ones nobody tells you outright—rules of survival.

No hot breakfast today, just a tin of ham and eggs (High Explosives the tankers call them), and the last of the salted peanuts from my care package. Tessa writes that another one is on the way. Hope it catches up with us soon.

0745:

We're on the move again looking for the next location to set up the Aid Station. Our jeep, ambulance and supply truck are tailing A company.

The Germans are fighting a rear-guard action as they retreat to the other side of the Maginot line. We're trying to move steady, just slow enough to keep out of range of their guns. We need to rest and regroup.

Exhausted tank drivers are falling asleep at the controls pretty regularly. They might find a ditch or just stop if they're lucky. We tape up some lacerations, and patch up more serious injuries as we run across them.

1420:

I set and splint up a driver's arm - radius and ulna broken, but not misaligned. Nothing penetrating the skin. He isn't in shock, so I give him a couple of aspirin for the pain. "Got anything stronger doc?" he asks. "You're being evac'd outta here," I tell him. "They'll wrap a proper cast around that arm and have something for the pain at the field hospital." I have a morphine syrette on me, but we're getting low. Besides Donne and Nicolosi got addicted to the stuff. "Doc, the pain's a son-of-a-bitch." I give him a saline injection and tell him "This'll make you feel better for now."

1535:

Corporal Wells from A company hotfoots his way back to our jeep and flags us down: "Captain, we found a cow laying in a field about two miles ahead. We haven't had fresh meat in a month. Mess Sgt. Kleckner hisself says he can slice it up for steaks. Ya know he was a butcher before he enlisted. We sure could use a meal like that. Can you give us the OK to serve it up sir?"

"Get in Corporal," I tell him. "Take us to this Grade A Prime."

1550:

A light snow whitewashes the field, shadows in shades of blue. For a moment, winter's beauty shines through, and I think back to Christmases at home.

Snap out of it, I tell myself. *There's work to do.*

Bossie is laying on her side, all dusted in white. Resembles a kid's sand sculpture at the shore. There's two craters in the field about ten yards from each other, and about forty yards from the cow. After brushing her off, no obvious wounds, not emaciated, not bloated. No evidence of a booby trap. Could've had a heart attack from the shelling. Nothing remarkable about her, but the face is sorry looking- big eyes open and glazed over, tongue hanging out, snot congealed on her nose. Maybe that's what a cow looks like when it's been scared to death.

I'm going by the book one more time. I thrust a knife through the shoulder until it hits bone, then withdraw the blade. No putrid odor, the wound doesn't splay open. Same result at the hip. Don't know how long she's been dead, but the cold acts like a meat locker.

Steaks would be a big morale booster. Food poisoning could take out more GI's that a German 88, at least temporarily. Might get a court martial, or maybe a demotion. The brass would find somebody to blame. "Go ahead corporal, she'll feed what's left of A company, and more, but I want to be there when she's butchered. And it's got to be cooked well done."

1640:

Nothing like fresh beef to bring out teamwork. Sgt. Kleckner cuts the cow's throat and secures her hind legs to a bracket and chain. Sgt. Rizzo commandeers a tank with a dozer blade and hoists the carcass off the ground. When I give the OK, Kleckner rips it from the rectum to the chest, then saws through the ribs on either side of the sternum. A few deft strokes of his blade, and the stomachs, intestines, bladder, hearts, lungs plop on the ground. A cursory look and sniff reveals no apparent indication of disease in the internal organs.

"OK Sergeant, cut'er up. But burn everything. Any food poisoning and you'll be on the hook. Understand?" He gives me a "Yes Sir." But, he's been around long enough to know that it's my ass on the line, not his.

1900:

The boys scrounge up enough dry wood to get a good fire going. No gasoline flavored chow from the stoves in the mess truck this time around.

Kleckner's a wizard with a knife. Name a cut, and its sizzling on the grate. Even the tongue and heart. Salt, pepper, fire, and hunger – all you need to make a dead cow into a feast for the gods.

Bottles of wine magically appear. Even Sergeant Mack managed to squirrel away half a case in the ambulance. Pudding and fruit cake from the ten in one rations finish up our fine dining.

Yes, enough food to fill bellies is a morale booster. It gives the men a chance to forget, even for just a few minutes, what they've been through, and what they'll face tomorrow.

So far, no indications of gastrointestinal distress, thank God.

One more chore for the night—filling out paperwork. To document today's feast, I report:

"Dead cow found in Field. Deemed unfit for human consumption."

At least the paper trail won't lead to me. Yes, I'm still learning the Army's unwritten rules. This one is called CYA.

*****

This story is based on actual events. They most likely did not occur in this exact sequence. The names are fictionalized.

Invader

By Janice Bilohlavek

IN LONG BEACH, CALIFORNIA around April 1965 the women's barracks had a fenced in yard behind the laundry room. Probably on a dare, a young sailor decided to climb the fence and enter the barracks through the laundry room. He was soon spotted and the warning rang out "Man on board." Soon doors started opening and WAVES came into the hall with brooms and mops or whatever they could get their hands on and started after him. He tried to escape by running up to the second floor as he did not want to have to pass the Chief's door. Alas the upstairs residents were also alerted and the poor fellow ran as fast as he could down the hall toward the fire escape. I am sure he was bruised quite a bit before he was able to get to the outside stairs. He hit the ground running and quickly disappeared never to be seen by us again. Lesson: Never underestimate the strength of a group of WAVES.

Toys in the Army

By John Steele

I WAS BLESSED with an incredible amount of toys that I was able to play with during my Army career. Half the fun of being in the Army was all the toys we got to play with.

Let's start with the vehicles. The first unit I was assigned to after basic had a bunch of 5-ton dump trucks. We spent our days riding around in the back of a 5-ton, bouncing down tank trails, hanging on for dear life so as to not be jolted off the back of the truck. Then they decided to give me a license for it. I had a one-hour class on how to do the PMCS (Preventive Maintenance Checks and Services) on a 5-ton, then I was told to get behind the wheel for a road test. The biggest thing I had ever driven at that point in my life was a station wagon. I drove the 5-ton around for less than five minutes, didn't run over anything, luckily, and they stamped my license saying I was certified to drive it anywhere, anytime. And they figured since I could drive a 5-ton, I could also drive a Deuce and a Half, a CUCV (Commercial Utility Cargo Vehicle), a M1008 (which was a pick-up), a M1009 (a Chevy Blazer or Dodge Ramcharger), and a 1/4-ton jeep, though only the First Sergeant ever drove our one and only jeep. Less than five minutes behind the wheel of a five ton, and the army automatically gave me a license for five other vehicles.

A few years later, they came out with the High Mobility Multipurpose Wheeled Vehicle (HMMWV) better known as a Humvee. We were told how they can climb a sixty percent incline, traverse a forty percent slope, and couldn't be tipped over nearly as easily as the old jeeps. All of that sounded like a challenge to us. Then they gave us the keys, not for the ignition, as no key is needed to start up the engine and drive a military Humvee, but to unlock the steering wheel, and told us to go play with them, see what they could do. For an entire day, we drove up and over every hill and rock we could find, and sure enough, try as we may, we couldn't flip them over.

Especially on warm summer days, with the top and the doors taken off, I did love driving the Humvees. I would volunteer to go up to Fort Drum sometimes to pick up supplies and such. One time I got pulled over in a Humvee on Interstate 81 on my way up to Fort Drum. The police officer asked me how fast I was going. I honestly didn't know. I was too busy enjoying the sunshine and the wind through my very short hair. The officer didn't know either. Military Humvees are painted with radar scattering paint, so he couldn't get a reading on his radar gun. But I was passing a car that was going sixty-seven miles per hour. The speed limit was fifty-five at the time. He said that he couldn't write me a ticket because he couldn't accurately say how fast I was going, but he could call my CO (Commanding Officer) and explain the situation, so slow it down. With that warning, he let me go.

I also got to drive a HEMTT (Heavy Expanded Mobility Tactical Truck), an eight-wheel drive 10-ton truck. An eight-wheel drive vehicle is obviously twice as cool as a four-wheel drive vehicle. The HEMTT weighs twenty-one tons. It provided good mobility over difficult terrain. One year we had an annual training up in a place called Deepwoods Maine, and as we were driving up there, someone hit a thousand-pound moose with a HEMTT. It killed the moose, but you couldn't tell the HEMTT was involved at all.

My favorite thing to drive in the Army though had to be a M113 Armored Personnel Carrier (APC). I was in a combat engineer company, and they were turning our unit into a mechanized engineer unit. For the APC, we spent about ten days learning the ins and outs and everything there was to know about the M113. But then came the day where we were turned loose with these 27,000-pound machines and were told to play with them and see what they could do. They could go up to about thirty-five miles per hour off-road, and could stop on a dime. You could stop so abruptly in this 27,000-pound vehicle at thirty-five miles per hour that the back end of the vehicle would lift a foot or two off the ground from all of its momentum. It could also carry eleven soldiers in the back of the APC. My soldiers loved it when they knew we were going to stop that quickly. They didn't like it so much when we didn't warn them.

I have to confess; I didn't drive the M113 all that much. I was a squad leader at that time; thus, I was the TC (Tank Commander). As a result, most of the time I got to ride around with my head and half my torso sticking out of the hatch. That was way more fun than driving it. But it was good and bad. Inside the APC, it was always warm. Or hot. On hot days, the rest of my squad was riding around inside a cramped metal box sweating their balls off while I enjoyed having half my body outside the hatch enjoying the breeze when we were moving. I would be caked in dirt because of all the dust and dirt that got kicked up on the tank trails, but I still loved it. On rainy days, they stayed dry while I still had half my body outside the hatch getting soaked. And in the winter, when the temperature was below zero and it was snowing, they were nice and cozy inside the warm APC while I still had half my body sticking out the hatch.

One fun thing with big, tracked vehicles is instead of going around objects, you can just go right over them, or through them. A bush is in front of you? Sure, you can turn and avoid it, but it is so much more satisfying to trudge right over it. There's a hedgerow up ahead? There's no need to detour, just full speed ahead and plow on through. A small tree? Why circumvent a tree when you can take it on, and probably win? With this mentality, we came up with a fun game. We decided to see how big of a tree we could knock over. So we had these 27,000-pound armored track vehicles going thirty-five miles per hour, ramming headlong into eight, ten, twelve-inch diameter trees. For the record, pine trees go down much easier than a maple or an oak. I am sure you've seen in movies where an APC or a tank drives over a car. I never got to do that. One of my regrets. Oh well.

For a few years, I was in a reserve unit that had a BEB (Bridge Erection Boat). The BEB had twin Sabre 6 Cylinder Inboard Turbo Diesel Engines, each engine with 255 horsepower. A couple times per year, we had to exercise the boat, so we would take it out on Irondequoit Bay and Lake Ontario. Those were good drill weekends, especially the first few times. But then members of our battalion headquarters came down once from Schenectady, and our Command Sergeant Major and Lieutenant Colonel Battalion Commander wanted to play on the boat. And they told their boss at brigade how much fun it was, and the next time we took the boat out, we had to let some full-bird Colonel from brigade play with the boat. Brigade in turn told division, and next time we had the boat out, we had some General who wanted to play with the boat. It got to the point where we no longer did training with the boat, instead we took a bunch of brass out on excursions so they could play on the water.

I am not sure if I should include helicopters as toys I've played with. I didn't get to play with them, as I was not allowed to touch the controls or fly them. But they sure were fun to fly in. Especially being that there were no doors on the ones I was in, and when they turned, they tipped some. There is quite an adrenaline rush when you turn your head to the right and see nothing but the sky above you, then turn your head left and see nothing but the ground several hundred feet below you, knowing that a secure seat belt that was made by the lowest bidder is the only thing preventing you from plummeting to earth. Somehow that was fun. Military cargo planes on the other hand were not quite so fun for me. Evasive maneuvers tipped the planes much like the helicopters, only you couldn't see out. That, and half the time, someone on the plane was barfing. It stunk, and the stench was nauseating. That took some of the fun out of the planes.

Another amazing toy that I got to play with were night-vision goggles. They are so cool to use, and it is something that everyday citizens rarely get the opportunity to experience. Again, night-vision goggles had to be used and tested and exercised. So one night, I had access to the vault, I took a couple pair of the goggles out and me and a buddy of mine took them to Durand Eastman Park. For several hours in the middle of the night, we walked the golf course and through the woods, checking out all the deer, raccoons, possums and other creatures that were roaming the park.

The final toy that I absolutely loved playing with is the EST-2000 Engagement Skills Trainer. The EST is used for weapons training. It is a computer-based weapons simulator.

It fires a laser at a huge screen, and you can see where the bullet would have hit. There are ten lanes, so ten soldiers can practice and fire at the same time. The screen is probably at least eight feet tall, and forty plus feet long. The laser weapon is the same size and weight as a real weapon, and when it is fired, it has the same recoil or kick that the weapon has. Same sound too, so it gets pretty loud. As for the weapons the EST can simulate, there are M-16 & M-4 rifles, M-203 grenade launchers, M-60 & M-240 machine guns, M-2 50 caliber machine guns, and 9mm pistols.

For soldiers who struggle with marksmanship, the EST is a great tool. The screen can present targets, and being computer based, the EST can help identify issues a soldier may be having while firing his or her weapon. And most importantly, it provides opportunity to practice without having to go to a range and expend ammunition. And it is safer.

Practicing marksmanship is good, but what makes the EST fun? Besides having ten lanes where soldiers are practicing marksmanship or simulating qualifying, there are two other modes of training that can be utilized. With ten lanes, you can have a squad with a couple of machine guns, maybe a grenade launcher, and the rest of the squad with their M-16's. Instead of having targets on the screen, you can have the screen showing enemy combatants attacking your positions like you are in a movie. What happens when you shoot at them? Well, if you hit them, they die and fall on the screen. If not, they keep coming. If they get too close, your part of the screen will flash red and your weapon stops working. You lose. The computer can send all sorts of scenarios at you, and can even adjust the number and efficiency of the enemy. The scenarios can put you in a jungle, then a desert or forest or grasslands or an urban area. It gets pretty intense, and it really helps a squad learn squad tactics.

Playing a simulation where you are getting attacked and you fight back is so much better than target practice, but even better is the "Shoot-Don't Shoot" scheme. People pop out on the screen, and you have a limited amount of time to decide to shoot or not shoot, and if you do shoot, you best be accurate. Simulating clearing a building or a house or walking down an urban street, can you make it through and remove all threats without taking out innocent civilians? It is much more difficult than you'd imagine. I was trained on how to operate the EST system, and on several occasions after the training day was over, I'd spend hours and hours at the EST building practicing under all the different conditions. It is the most expensive and realistic video game I've ever seen.

* * * * *

What you are reading now is an amendment to my original story. You probably noticed that the words are now italicized. How did I manage to leave out one of the most fun things? In the Army, I was a Combat Engineer. Combat Engineers get to play with explosives. Can you say "trinitrotoluene"? I can't. I had to look it up. But it is what TNT is short for. Numerous times in my Army career, I got to play with TNT and C-4 and det cord (detonation cord). Part of my job was to blow things up. For a teenage boy, you would be hard pressed to find anything as fun, thrilling or entertaining as blowing things up, with the possible exception of being with a teen-aged girl.

I was also advised that my piece ended abruptly, it needed some type of wrap-up. To tie it all together, I think everything I played with in my Army career should be featured in an Army recruiting video. Driving vehicles that cost hundreds of thousands of dollars each, ramming those expensive vehicles into and knocking over trees or running over other inanimate objects. Playing with intensive, realistic, immersive, state of the art video games. Blowing things up. I think we have a winner. Young people will be clamoring to join up and play.

Carry a Gun in Nowhere

By Lee Nelson

GUNS ARE A RATHER OBLIGATORY PART OF THE MILITARY. If you serve your country, you're gonna see one or two guns, probably even fire one once or twice. I can speak for an Air Force experience, and everyone receives hands-on training with at least the M16 in basic training. Depending on your military profession, the M16 may be the tip of the iceberg, but I wouldn't know. It certainly wasn't for me. You don't see too many guns cleaning teeth, which is largely what I did for my four years of service.

However, my four years of service ended up proving to me, ironically, that I didn't need the military for a life of hearty gun exposure. All I needed was to be surrounded by avid gun enthusiasts. You don't need the military for this, but the military provides gun enthusiasts anywhere you move. I can think of maybe a handful of people I met in the military that didn't own a personal weapon, myself included. You couldn't find more gun enthusiasts if you moved to Montana. Anyway, you're gonna see guns in the military, lots of them. All you need is friends. If you don't have the guns, they do.

Some of these friends might come to you with a problem with their guns. I know mine did. A big problem with personal weapons in the military is where to store them. You see, the military isn't too keen on uncontrolled weaponry. They're really big on having all the control in this department, especially if you live on base. So, while this is a damn reasonable expectation on paper, look at it this way: the military demands its gun toting military to surrender its weapons. Yeah. Imagine that. Try not to laugh while getting your head around this. Many of my fellow servicemen couldn't tell you there's a Constitution if not for the 2nd Amendment, so do you think they were going to surrender their weapons? Nah. No. No active duty military members are ever going to surrender their personal firearms. This was never going to happen, and to this day, I assure you, this is never going to happen.

So, while I never owned a gun, I had the good fortune of being the married guy who lived off base. If you're this guy, your single, gun toting friends will pursue you, and you'll be happy to store their guns for them. Seriously, you're in the military. You're gonna tell me you have a problem with this? I had lots of guns in my off base house, lots and lots of guns. I had big guns and little guns, hunting rifles and hand cannons. You name it. I'm gonna use assumed names to protect the guilty, because quite honestly, not all of them were legal.

So, I had these friends, Billy, Bob, and Eddie. They really dug my quiet custody of their weaponry, and they always invited me on excursions with their large guns, and you know what? Guns can be a lot of fun. Everyone should try a gun once or twice.

Billy, Bob, Eddie and I were stationed in the middle of nowhere. To give it a name: Minot, North Dakota. There are other military middles of nowhere, and this one was ours. On one of our fun gun excursions, we drove to absolutely, positively the middle of nowhere, and fired obscenely powerful weaponry into about 20 cases of empty beer bottles, and several unfortunate prairie dogs, and called it a day. We then proceeded to get stuck. I recall we were concerned about the mud out there in nowhere, but Billy had a four wheel drive Jeep. Freaking Jeep, right? What's a little mud?

A lot, apparently. You never see war movies where the Jeep gets stuck in the mud, but there we were, and no four guys with all the guns in the world could get this damn Jeep out of the mud. We tried all the tricks four men from four different parts of the country could pull out of their asses. No such luck. We left on foot with our infantry of weapons in search of the nearest neighbor that would help. Why not, right? One should arm oneself in nowhere. What else is the 2nd Amendment good for?

Before too long, we found a rather massive farm that appeared to own Nowhere. It had a red barn, the largest red barn I've ever seen. It contained two giant combines, tons of expensive and specific looking hard labor equipment items, an incredible tool collection, and a phone. Very high end barn, but who owned it? Who owned Nowhere? There wasn't a human being to be found for a thousand miles. We didn't want to just use their phone. So, we rounded this massive red barn (with our guns, of course), and looked in all directions. Finally, there was a sign of life, and we all saw this sign of life at the same time. It was a striking sign of life.

Eddie was the first to react. He turned and looked at all of us and whispered very adamantly, "Not a fucking word!" Yeah, it's not every day four horny kids find a naked, sun bathing trophy wife in the middle of absolutely nowhere. And there we were with her. There she was like some beautiful mirage of The Great Plains. We couldn't believe our eyes.

And nothing else happened for quite some time. We really didn't want to scare her with our guns, and we really didn't want to, quite honestly, make her move at all, or do anything unnatural simply because we were too stupid to get our Jeep out of the mud. So....we had to think about this. How could this be properly handled?

Eddie was prepared to think indefinitely. The trophy wife did a lot for his focus. Bob and I were quietly giddy. Billy was strangely (and annoyingly) concerned for the welfare of his Jeep, so things eventually came to a head. Bob suggested we go back inside the barn and make a lot of noise, shouts of hello. I suggested we lay our guns down before we do this. Eddie offered nothing. Billy was down with the plan. We stood the guns against a wall inside the barn. Still, we took a little extra time before committing to any of this. Eventually, we made a lot of noise in the barn while Eddie stood guard and waited for a reaction.

Eddie was a statue peaking around the corner. The Honor Guard didn't have shit on him. He said she vaporized into the house. Not much of a show. Literally .2 seconds later, this cute little Houdini rounded the other side of the barn in jeans and a t-shirt, and zero congeniality.

She asked us what the fuck we wanted, and informed us how soon her husband would be home. Then she noticed the guns. What's with the guns? That was a good question. Four guys with guns might be a touch off putting in Nowhere, and you have to have grit and fortitude to live in Nowhere. This little cutie might just be the resourceful type. Does she have her own gun in those jeans? Smart damsels don't allow for distress. Maybe we should all pretend there are no guns in this barn and act natural, be friendly, and nobody will fear for their life. Nobody will fear getting kidnapped, nobody will fear looking like a kidnapper, and maybe no farmer will end up tilling us into his grain elevator. Quick words were required.

And out of nowhere, Eddie had them! He saved the day! I could tell other stories where he was good for that, and Eddie explained every detail of our mess to her satisfaction. Time passed, and her husband finally arrived. He was a large and somewhat grizzled gentleman, and he brought his own gun to the party. Like I said, bring a gun to Nowhere. It's wise, and as the situation progressed, even good etiquette. He liked our guns. He liked them a lot. He was full of questions. The man knew his guns. With the guns in common, he was far more amused with the situation than his trophy wife. After some chatter, he collected us and our guns into his truck, brought us to Billy's Jeep, and pulled us from the mud. All ended well. We still had time that day to hit a couple of bars. We offered him a few beers for his troubles, but he declined.

Oh, and for the record, Eddie didn't exactly explain every detail. He edited for content. He kind of left out the part about the time spent around the barn, and any admissions that we witnessed this poor woman's nakedness. He was a good editor. He saved our asses and her dignity. I've often thought of Eddie over the years. Whatever became of him? I imagine him negotiating hostage crises at large.

Needless to say, this wasn't the end of our frontier gun adventures. We learned to never get stuck in the mud again, but North Dakota is built for exactly those shenanigans. Those open plains don't offer much else, and those prairie dogs probably still speak of us at campfires. Our ancestral legend will live forever, one terrified prairie dog pup at a time. Finish your meals, children. Mind your manners, or...

What can you say? Young men get bored and do young men things when they're bored and unsupervised. We could've done worse. The prairie dogs may beg to differ, but we could've done worse.

Actually, we did do worse, but this was definitely the only occasion when the innocent were victimized.

And we all turned out okay. The guns may have had nothing to do with it, but the military sure did..

Some Navy Memories

By John Michael (Mike) Ryan

1 Editor at Large

IT WAS 1971, and my job as Lithographer aboard the Light Missile Cruiser Springfield presented me with an unusual opportunity to break up the standard content of The Sea Dog, the ship's newspaper.

Routinely, the content consisted of the POD (Plan Of The Day), world news, biography of the Seaman of the Week, and the comics. The latter was probably the first read of the day for the crew of 1200.

Sporadically, classified material was received, already in writing. I had to make extra copies in the shop for distribution to our Vice Admiral and to the 200 officers on board. Naturally, the "swallow or burn after reading" warning was assumed. Burning was the choice.

I was known for my quiet demeanor and dedication to my job. However, as the one-man night crew in the print shop, I fashioned a different way to notify the crew of my impending transfer to the Naval Photographic Center in D. C. I inserted a photo of my handsome self into the Dagwood comic of the day. Reaction was immediate. Our Legal Officer sought me out with (tongue in cheek) threats to scuttle my transfer and to bring me up on charges of misuse of government property and changing the ship's protocol without permission. Thankfully, no Brig time occurred. After I had left the ship, I was voted Seaman of the week and was written up in a very nice article in The Sea Dog along with a photo of me at work.

2 Quick! Follow that Tray and its Contents

The Springfield is a good sized ship. It measures 66 ft wide and over two football fields long (610 ft). It is known for being stable in high seas. That reputation was easy to buy into except, sometimes, at meal time. One day, there were fewer in the chow line because of the turbulence going on outside and the rolling of the ship. I was seated at the middle of a bolted down dining table. Three obvious rookies were in a row nearby. I watched them eating. They had a utensil in one hand with the other hand on their side. Their feet were resting lightly on the deck. There was no problem, there being a temporary pause in the sea swells. I sat with my forearms pressing down on the table with just enough room between them to block my tray if necessary. My feet were pressing down on the deck. I made short, quick sweeps with my utensils. I realize now, I eat that way today.

Suddenly, the ship listed a good 20 degrees. My tray went nowhere but the rookies, in their chairs, slid about 25 ft and crashed against the bulkhead. They were in a tangled heap on the floor next to their well traveled food and drinks. There was no reaction nor conversation. Maybe they were in shock.

I finished my meal and went back to work using my skillful, rocky seas stride. Or was it my rocky seas shuffle?

3 A Furry Candy Thief

While off duty in the Mediterranean Sea, a buddy and I went ashore to Gibraltar to pass some time. There we were greeted by Barbary Macaques Apes. They reside in Gibraltar's Nature Reserve but some of the 200-300 residents venture out into the public. Barbary Macaques Apes are only about 24 inches long. Adult males weigh around 35 lbs, females around 20 lbs. Both have large, sharp teeth. They eat plants and insects. They also can distinguish, somewhat, between red, green and blue. The story continues with me taking a photograph of my crew mate while he was sitting on a large rock, facing me. There was a sign posted to his right which read: "Do Not Feed the Roaming Apes." Are they referring to the inhabitants, my buddy, or to me? We reversed roles with him taking a picture of me squatting ten feet in front of him. I was holding a green wrapped piece of candy. An Ape sauntered up to me and snatched the candy from my hand. One look at those teeth convinced me to not argue.

4 An Ice Cream Offer and a Camera Mishap

The Springfield was the flagship of the 6th Fleet. The Leningrad was the flagship for the Russian's Mediterranean fleet. Our missions were the same. We patrolled that big pond in times of stress and in times of peace. A peaceful meeting of the two did take place while I was on the Springfield. I was able to see much of what happened from my print shop hatch (door). Other details came from my crew mates.

The Leningrad pulled alongside the Springfield no more than 60 yards away. For some reason, they had some crewmen at attention on the deck, in dress uniforms and facing us. We were walking around in our work clothes. Both ships had an attached helicopter with a landing pad. Their helicopter was flying back and forth between the two ships. One guess was that they were giving us a greeting and showing respect. Another theory was that it was a ruse and probably a reconnaissance ploy.

Our response was to signal them this message; "Do you want some ice cream." They declined. Nevertheless we dispatched our Sikorsky HC-4. Det 41 helicopter to their ship. As it hovered above their fantail we signaled another offer of ice cream. Again, they declined. Our crew returned to ship. In the meantime, their chopper was still doing its thing with a crew member trying to take photographs of us, up close and personal, so to speak. He was holding a huge camera. My guess it was a Bronica SQ-A, Medium or the Russian equivalent. As the photographer tried to remove the lens, he lost his grip and the lens went into the drink. I hope he wasn't docked for the lens.

Both ships returned to their normal routines. Party time over!

Hurry Up and Wait

By Sue Spitulnik

ONE OF THE MOST COMMON PHRASES I remember from my years as a military wife, 1972 -1979, was "hurry up and wait."

Back then, when someone enlisted, it could be weeks before they went to boot camp. Five years ago, a family friend signed on the dotted line and didn't start his service for six months. We need you, but not quite yet. Amazing.

Anyway, it quickly became the norm to be told to make a decision quickly but then have to wait for the repercussions or answers to that decision for much longer than you had to make the choice. Take, for instance, requesting what bases you wanted to "put in for," compared to "you will be assigned to." Remember, I'm trying to remember back fifty years. My active duty member came home with papers to fill out regarding his choice of an overseas duty station. There were no computers then, and most of the guys followed what their higher-ranked friends advised." Never put your first choice first; you will not get it." So we discussed it as a couple and filled in the blanks with our requests. As the guys said, we got choice two: the United Kingdom. At the time, there were over ten bases in the UK. Months later, we finally moved and were stationed at RAF Lakenheath, home to the one military hospital that cared for all the American service people and their families stationed in England.

I was pregnant with our second child and soon found the busiest clinic in the hospital was the OB/GYN clinic. They insisted you be on time for your appointment, but they scheduled four ladies for the same time in fifteen-minute increments. Sitting in a packed waiting room was common. There would be the first-trimester women, glowing with the fact of carrying a new life. The second-trimester group was still comfortable baby-weight-wise. Everyone was gabbing about names, where they called home, and how many times they had been pregnant. Then there were the third-trimester ladies. Sometimes a little puffy, wiggling in their seats, trying to sit comfortably and wondering what they were doing there every week to get their belly measured, a heartbeat listened to, and hoping not to have to get undressed in the exam room. Back in the 70s, there weren't ultrasounds, so no one knew the sex of their child ahead of time, nor were they aware of now predictable problems, and due dates were a good guess. I can say this because I was there. We resembled a herd of cows waiting to get milked so we could get on with our day.

My daughter was born in December 1974. She immediately taught me more about hurry up and wait when she wasn't able to tolerate my milk and started losing weight in the nursery. Finding the correct formula for her took a while, so I went home ten days before we could bring her home. When she was about a year old, she started having sudden onset high fevers. Off to the hospital we would go as it was an all-encompassing building. The nurses and doctor would do tests looking for a cause. When they drew blood, they sent me into the hall saying, "We'll bring her out to you when we are done, and you can comfort her. She'll be mad at us." It worked. Then, the urine test had to be accomplished. They would attach a small receptacle bag with a sticky rim to her private area. It looked like a miniature vacuum cleaner bag. Then, the waiting for her to go began. I'm sure it wasn't comfortable for a little one, especially when I couldn't explain things to her, so she held her urine, sometimes for over an hour. Of course, she was fretful and didn't feel good because of the fever, so her Dad or I would carry her around the hallways, look at pictures, give her more to drink if she would take it, and take her into a bathroom and turn the water on. Eventually, she would urinate. I don't remember her fussing when the bag was removed, just relaxing with a sigh that the torture was over. Each time, the urine sample would prove she had an infection. After a round of antibiotics, she would be good for a while, and then we would repeat the process. And each time she had to have that bag attached, she held her urine longer.

When we would go in for a check-up having to do with urinary tract problems, she often had to have her blood drawn. She got to know the hospital hallways, and as soon as we turned the corner to the lab, she would start screaming. Heads would pop out of office doors like the critters in the "Whack-a-mole" game, looking to see what was wrong. As soon as I said, "She knows where she's going," they smiled with understanding and disappeared. She stayed mad at me for letting people hurt her after those visits.

Just before she was two, she and her brother were chasing each other from bedroom to bedroom, and she slipped on a piece of clothing and fell. The screams were instant. My mother had taught me to never pick up a hurt child, to let them get up and come to you. I crouched beside her and held out my arms. She sat there and cried louder. I had the neighbor drive us to the hospital. After x-rays showed her femur was broken, they told me they would apply a unilateral bilateral cast because to immobilize one hip, you have to immobilize both. I had no idea that meant a body cast with a bar fixed between the two knees, so there was a way to carry her flat. The back of the Spica cast was rounded out over her little bum. Children are resourceful, and after a couple of weeks, she learned to use her feet and shoulders to flip herself over like a turtle. She ended up shattering that back-raised edge, but the pieces of plaster stayed inside the wrappings for the last week she had it on. I still threaten her that I will share the photos of her during that period with anyone she marries.

We found out later her femur breaking was related to her kidney problems. When she was two and a half, after a base change to Rantoul, IL, she had her urethras re-implanted into her kidneys on each side. After the surgery, she had a catheter for a couple of days. We would visit her in the hospital and walk the halls so she could be out of bed. It felt like walking a dog on a leash, but the difference was there was no pulling on the leash. She picked the way we walked and at what speed. And she's still doing it today.

There are other stories of "hurry up and wait," but not as unique as these.

Beer Adventure in Saigon

Gary Redlinski

HAVING SOME TIME OFF from our assignments at the US Army Mortuary on Tan Son Nhut Airbase, in Vietnam, probably, in 1970, Bob K. and I wandered into downtown Saigon. We stopped at the Continental Palace Hotel. The Hotel was an upscale place. It had a veranda with a stone railing, where you could eat or drink in a comfortable setting.

Sometimes being a seasoned veteran, after 18 months in a war zone, can make you a bit complacent. Especially when it seems like a normal day in a peaceful country.

We sat down near the railing to have an unobstructed view of the city street. We ordered a couple of beers and were relaxing and enjoying the day. It was a beautiful sunny day, and the traffic was fairly light for that time of day. As we were talking and enjoying the view, we heard some "popping sounds"- gunfire. We looked around wondering where the shots being fired came from. Suddenly 3 or 4 young guys came running down the street. Shortly after came the "White Mice," (what we affectionately called the local Vietnamese Police.) They wore white and gray uniforms and all their holsters and belts were white. Their guns were drawn, firing at the two running guys.

To say the least, Bob and I decided we could get hit with a stray bullet or two, sitting next to the railing. Without saying a word to each other, we quickly knew we shouldn't be sitting in the open. We quickly grabbed our beers and moved to a table closer to the building so we would be less visible, and harder to hit. By that time they had run past us, so we continued our refreshments from a safer distance.

After the excitement was over, and a few more cold beers, we hailed down a local cycilo (sort of a rickshaw, but with the driver in the back, and you are the 'bumper') driver and headed back to the airbase.

All in all, it was a fun time off from our mortuary duties.

The Specialist

By Tim Hansen

THE MECHANICS AND DRIVERS always amused me when I was the battalion motor officer for Second Battalion, 10th Special Forces (Airborne) at Fort Devens, Massachusetts in 1989. The monthly Unit Status Review, the routine 10 percent inventories, and the weekly preventive maintenance, checks, and services, or PMCS, filled my calendars and taught me a critical part of logistics. But, it was those guys in their olive-drab green coveralls, faces covered with sweat and grease, whom I remember most.

I particularly remember the time when Sergeant Dennis Bradley spent time preparing one of his E4s to stand before the board for the Platoon Leader's Development Course. Specialist Joe Eagleton was a light-wheeled vehicle mechanic from Binghamton, New York, and at 32, he still had a boyish look. Bradley spent hours with him reviewing common soldier skills, leadership, drill and ceremony, and physical fitness. Eagleton admittedly wasn't the sharpest wrench in the toolbox, but he was honest and always spoke his mind. Most of the time, he drove Bradley nuts with his unique logic in repairing Army vehicles, but for the most part, the two got along like two brothers. When the day of the board arrived, Bradley had Eagleton looking perfect in his Dress A's. When I stopped by the garage later that day, I noticed Bradley sitting at his desk, reviewing work orders and seemed far removed from everyone. In a mood, I thought. I walked out to the bay to check the latest arrival of repair parts and noticed Eagleton hard at work on a two-and-a-half-ton truck. I walked up to him as he repositioned a wheel back on the truck.

"How did the board go?" I asked.

"Oh, pretty well, sir. I got most of the questions, but then the sergeant major threw one at me I didn't expect," he replied.

"Oh, what was that, Eagleton?" I countered.

"Well, he asked me what I thought my next career move would be," he replied.

"Sounds like a fair question," I said. "What did you tell him?"

"I told him I thought the Air Force looked pretty good," Eagleton said.

"Air Force?" I said incredulously. "I thought you were shooting for PLDC?"

"I am, but I said the first thing that came to my mind."

"So, is that why Sgt. Bradley is sitting by his desk, not talking to anyone right now?"

"He's not talking to anyone for the rest of the week," Eagleton replied sheepishly.

"Well, Eagleton, chalk it up to experience and try again in six months," I said.

Eagleton shook his head and went back to working on his truck.

An Epicurean Disaster

By Chris Cilento

"YO, SGT C! HEADS UP!"

I looked up from my rifle to see a small brown package careening toward my head. Instinctively, I ducked and batted it away with my hand and watched as it fell to the ground.

"Good thing that wasn't a grenade, Sarge! You'd have just blown us all to hell!" With a derisive snort, I looked at the faded brown plastic package at my feet. The lettering said "MRE." I picked it up and muttered, "might as well be! After a few days of eating these the results can be catastrophic!" Squinting in the bright desert sun, I looked closer at the lettering.

"Spicy Pasta. Vegetarian." I grunted and glared at the man who threw it. Holding it up I said "Sanders, are you tryin' to tell me something?" His toothy grin shone white through the dusty haze.

"Yeah! Get off your fat ass and join us in the gym once in a while!"

"Well maybe if you clowns would stop getting into trouble I wouldn't be stuck filling out so much damn paperwork! I still owe the commander a report about that Lieutenant from Bagram!"

Sanders's grin faded to a scowl. "I told you I didn't know she was an officer! Damn if she didn't have a mean right hook though!"

"Serves your sorry butt right," I laughed. "You're damn lucky she didn't press charges."

Sanders grinned again. "Well at least I've got a story for the grandkids!"

I couldn't help but grin as I shook my head. "You're trouble, y'know that? You've been trouble since I first laid eyes on your ugly mug at MEPS. How the hell you managed to make it through basic I'll never know."

"My roguish good looks, my friend! Gets the ladies every time!"

"You realize all our drill sergeants were dudes, right?"

Sanders struck a pose and said, "Hey, I can't help who finds me attractive! You're just jealous!" I laughed and chucked a rock at my longtime friend. "You're an idiot, y'know that, right?"

With a sigh, I leaned my rifle against the withered tree which provided a meager shade and pulled the MRE package open. It was the thing I hated most about being in Afghanistan. If the insurgents didn't kill you, the food would. We had been on patrol for four days and were looking forward to returning to our FOB in the morning. Two or three times a day we would choke down what can only be described as dog shit with enough sodium to kill an elephant. Whoever decided this garbage was edible should be dragged out into the street and publicly humiliated. Four days of nothing but MREs and warm, stale water. How long until my enlistment is up?

Opening the "spicy pasta" I almost gagged at the smell. With a sigh, I rifled through the "extras" bag looking for something, anything to make this mess the least bit palatable. Salt and pepper. What kind of idiot loads these things up with sodium and then adds salt packets? I looked up and called out to the rest of my squad "anybody got any hot..."

Suddenly I was pelted with half a dozen tiny bottles of hot sauce. My guys knew I hated MREs and the hot sauce, if not making them edible, at least burned off your tastebuds so you didn't have to suffer the taste. Chuckling, I picked them up and immediately pocketed three of them for later. "Thanks, ya bunch of lowlifes!"

Private Cornelli, the youngest of my soldiers piped up, "Awww we love you too, Sarge!"

Sanders hollered back, "Shut up, Carl!"

Opening three of the bottles of hot sauce, I dumped the contents into the open MRE bag. If it wasn't spicy before, it sure as hell would be now. With a long sigh, I dug in. Better to get it down quick than have to suffer longer. The smelly concoction coated my tongue with a slimy, salty film. An aftertaste like rotten skunk made my eyes water, or maybe that was the hot sauce, as I shoveled the foul stuff down with a false bravado. I could feel it sliding down my throat and dropping into my stomach like a depth charge. As I ate, I swore vengeance upon the makers of this rubbish.

Sanders wandered over later with a cup of something black and steaming. "Here ya go, Battle! Some fresh coffee for you."

Eyebrow raised, I looked up at him, "Fresh? Maybe. Coffee? In a pig's eye! Gimme that. At least it has caffeine."

In mock indignation Sanders shoved the hot metal cup in my hands. "Look, just because you're some fancy, dancy epicurean cuisine master at home doesn't mean you can just shit on my efforts to provide your sorry ass with sustenance."

Both eyebrows went up this time. "Wow, look at you, Sanders! Been reading a dictionary lately? Not like you to use all those big words!"

He punched my arm and jostled the hot liquid. "You're an asshole!"

"And you're not?" I grinned, taking a sip of the acrid liquid. I winced at the bitter flavor but found a small measure of comfort in the familiar flavor. Army coffee was definitely an acquired taste.

"At least I own my shit!" Sanders retorted. "Come on, Battle. Let's go check on the guys."

Grasping his outstretched hand, I grunted as he helped pull me to my feet. "Jesus, man! You really gotta get to the gym more!"

His sarcasm was met with a stony gaze and a gesture inviting ornithological discussion. My gesture was returned with a hearty laugh as we walked over to where the rest of our squad was trying their best to enjoy their evening meal.

It had become a tradition with us to make sure our squad was doing okay. Every evening we would check on each soldier, ensuring their gear was good, they were physically alright, and evaluating their mental state. War has a way of taking its toll on even the most resilient soldier.

Private Cornelli was fresh out of basic training and only 18 years old. He was the most gung ho of all my soldiers but he had not been prepared for the rigors of battle. Some of the things we had seen had left their mark on him. His humor and enthusiasm were coping mechanisms but I could see the haunted look in his eyes. I sat next to him and gave his shoulder a good-natured punch. "How's the chow, Cornelli?"

He snorted and looked over at me, "Tastes like rotten camel crap, Sarge!"

"Oh, so you got one of the good ones then!" I said with a grin.

He laughed but was suddenly quiet. "Ain't nothing like my Ma's cooking. I'd give my left arm for her stuffed shells."

With a sympathetic nod, I said "I feel ya. I'd give real money for a nice juicy steak with all the trimmings."

Cornelli had a far away look in his eyes. "Remember that Christmas dinner you made for all of us before we left? That was a good time. I don't think we've had a good time like that since."

"Yeah," I said. "That was great. I enjoyed cooking for you guys."

Sanders was making his way back to where I was sitting having already checked on the other three soldiers. He saw the look on Cornelli's face and glanced at me. I nodded and gave him a knowing look. It's amazing how much information can be passed with just a look. I made a mental note to make sure Cornelli got first dibs on the phones in the USO so he could call his parents tomorrow. Having been raised in a large Italian family myself, I knew how close knit they could be and how much it hurt to be away.

The next morning, we made our rendezvous point and were picked up by two Blackhawk transport helicopters. Leaning into the webbing of the seat I closed my eyes and relaxed for the first time in days. My mind wandered to another world in another time. A world where dirt and dust were not a constant, where clean was not a metaphor, where food was something to be enjoyed.

The Christmas before I deployed I had hosted a dinner party for my squad and their families. It was a classic Christmas dinner with all the trimmings. I found a turkey of Dickensonian proportions and deep fried it. The hot oil instantly seared the skin and kept the juices from escaping. It was the juiciest bird I had ever made. Real mashed potatoes with giblet gravy, sweet potato casserole, hand churned butter, homemade cranberry sauce, and scratch made dinner rolls;

fresh sautéed green beans, steamed broccoli with cheddar cheese, and real, homemade apple pie and hand mixed ice cream.

The tables groaned with the weight of the feast. I spent a solid week cooking and preparing and it was worth every second. For a group of people used to chow hall food, it was a culinary experience that they would never forget. I remembered the look on their faces as they each came inside. My house was decorated to the hilt and I had found a tree so large it touched the ceiling of my living room. Candles gave the house a warm and inviting look and the smell of allspice, pine, and cinnamon, combined with the aroma of the huge turkey, was intoxicating. They each gazed wide-eyed at the spread and sniffed the air with excitement. That was the last time I could remember feeling happy and content.

With a start, I was awakened from my reverie by Sanders as he slapped my knee. "Whatcha' dreaming about, Battle? Must be something good with that smile on your face!"

"You remember the Christmas dinner I made just before we deployed?" I asked. His eyes got a distant look as he remembered the event.

"Dude, I can still taste that turkey! The guys still talk about that dinner! What the hell did you do to make it so damn good?"

Grinning, I just sat back and closed my eyes again. "A good chef never reveals his secrets!"

After arriving back at FOB Shank, I sent my guys to stow their gear and get a much-needed shower while I made my report to the Commander. After cleaning up, myself, I met my squad at the chow hall. It was a Friday and we had smelled the enticing aroma of steaks cooking on the grill as soon as we landed. We were looking forward to the once a week treat of a surf and turf dinner. It wasn't Delmonico's but it was a damn sight better than MREs.

As we moved through the line, I noticed that there had been a change in management at the chow hall. The new company was called Sodexo and the servers and managers were all in bright white jackets and tall paper toques. Fancy, I thought with a shrug. Chow halls changed hands so often here that it just became a matter of course. Grabbing a tray, I bellied up to the serving line, anticipating a nice juicy steak, shrimp, and lobster tail after a grueling 4-day patrol. I was utterly dismayed when I saw what was being served. What once may have been a nicely grilled steak was now sitting in a vat of greasy water. It hit my plate with an oily "slap," splattering noxious liquid everywhere. Looking in shock at the server I exclaimed, "What the hell is this?"

"Steak, sir!"

Looking askance at the ruined slab of meat I said, "What the hell did you do to it? Why's it sitting in water?"

"Keeps it juicy, sir!"

"You've ruined a perfectly good steak! What the hell is wrong with you people?" Grabbing my plate in disgust I moved down the line. The lobster looked like it had been rescued from a desert island and was tough as shoe leather. The shrimp...well, if they were trying to make an alternative to bullets they succeeded. I voiced my complaints at each station, berating those that would dare to serve such tripe. Angrily, I stomped over to the table with the rest of my squad. I looked in horror as, to a man, they ate with gusto; clearly enjoying their meal.

"How can you enjoy this garbage?" I asked in disgust. "You're all barbarians."

"*Awww* calm down, Sarge! Not all of us grew up in a cultured home!" Specialist Gomez chortled, elbowing his buddy. Our resident loudmouth and jocular smartass continued, "the way you act, you'd think they did it on purpose just to piss you off!" Glaring at his insolence I sat down and tried to ascertain which mess I would try to eat first. I grabbed a bottle of steak sauce and drowned the wet slab of meat in the spicy substance. It wasn't like it could be saved anyway. To my distinct displeasure the meat was not only wet and mealy but tough as an old boot. Disgusted I chewed in silence, mourning what could have been a decent meal and shaking my head at the obvious enjoyment my troops were showing at the horrible repast.

Suddenly, a shadow loomed over me. I looked up... and up, and up. A giant of a man stood before me. His stained white jacket stretched to the breaking point over bulging muscles. He loudly exclaimed, "My servers say you don't like the food! You got an issue, Sergeant?"

Slowly putting my fork down, I stood up. Out of the corner of my eye I saw Sanders elbow one of our soldiers and say, "Watch this!" I could feel the eyes of everyone in the chow hall watching me. My eyes came to the level of his enormous pecs, his nipples almost large enough to poke my eyes out. I thought, *could this guy be any more cliché?*

Planting my feet and refusing to be intimidated I said, "And you are?"

His dark face came inches from mine. I could smell the garlic on his breath. "I'm the head chef. You got a problem with the food you talk to me!"

"So, you're the one responsible for this gastronomic disaster?" I glanced at his nametag,

"Dwayne, is it? How can you call yourself a chef and serve this kind of garbage!" His eyes widened at the insult.

"I'll have you know I studied at the Cordon Bleu! If you don't like the food feel free to pick up an MRE on your way out!"

Glancing skeptically at his dirty uniform I said "You studied at Le Cordon Bleu? I find that incredibly difficult to believe. Any Cordon Bleu chef would be ashamed to serve garbage like this and wouldn't be caught dead in a dirty uniform. Maybe you should consider some remedial training?"

Dwayne leaned even closer and grabbed the front of my uniform with hands the size of a skillet. The reek of his breath burned my eyelashes to ashes. "Now listen here you little pissant! Nobody comes into my kitchen and insults..."

That was as far as he got before my entire squad stood up and surrounded him. Sanders tapped him on the shoulder and said "Sir, I highly suggest you take your meaty paws off my sergeant and go back to your beloved kitchen. Otherwise, the chow hall will be looking for a new head chef. Now I'm sure SGT C. didn't mean to insult you, personally. It's just that we've been on patrol for 4 days and he's not a real fan of MREs. He's a bit of a food snob, really, so you'll have to forgive his indignation. Personally, I thought the steak a fine example of grilled boot leather. The lobster was over-cooked flawlessly and I never expected to get blackened shrimp in Afghanistan. Go on back and keep up the work you've been doing. One man's opinion really isn't worth getting the shit kicked out of you." He patted him on the back in a friendly manner but there was ice in his eyes.

The chef stood up, looked around, and focused his attention on my friend. "I suggest you and your buddies leave my dining facility right now."

Cheerily, Sanders slapped the giant man on the shoulder. "Sure thing, buddy! Soon as we've finished our meal."

The big man clenched his huge fists and stood nose to nose with Sanders and growled "You'll leave now or you'll be taking your meals through a straw!"

Six pairs of fists came up. The silence was so thick in the chow hall you could have heard a mouse fart. I noticed four MPs step into the large room and knew it was about to get real.

"Alright, alright!" I said. "We're leaving. *D-man*, you mind if I call you that? I apologize for insulting your esteemed education. Clearly you know far more about cooking shoe leather than I do. Sanders, take the boys back to our hooch. I'll be there shortly."

"But Sarge!" one of the Privates groaned, "We haven't finished!"

"NOW!" I exclaimed.

Sanders, always quick to *suss up* a situation, followed my glance toward the MPs and said "You heard the Sarge, guys. Let's go."

He gave the chef another friendly slap on the arm and said, "Next time, big guy!"

As they hurried out of the building the MPs walked up. "What's going on here, Sergeant?"

Since the man outranked me, I assumed the position of parade rest and said, "Not a thing, Sergeant! Dwayne and I here were just discussing the finer points of the culinary arts. He studied at Le Cordon Bleu and I'm a bit of an epicurean myself. We were just... trading recipes."

The MP looked at the giant chef with suspicion. "Is that right?"

Dwayne looked down at me. I raised an eyebrow and shrugged.

"Yeah, that's right," he muttered and stomped off toward the kitchen.

The MPs shrugged and walked off. I breathed a sigh of relief. "One of these days your mouth is going to get you killed, you know that, right?" I said to myself as I gathered my tray and dumped it into the trash can.

Walking back to my tent I dreamed of a day, not incredibly far off, when I would get on a plane and leave this god-forsaken country and its abominable food behind. Remembering how happy it made my guys before, I decided I would throw a celebratory feast for them when we got home. It would make the Christmas dinner pale in comparison and would likely cost me most of my accumulated combat pay. But some things were worth it.

Do Not Put Your Finger In This Crack

By John Michael (Mike) Ryan

HERE IS THE BACKSTORY to a tale of a shakedown cruise that went sideways. I was nineteen years old, having joined the Navy in August of 1969. My new floating home, the USS Springfield CLG7 was docked at Norfolk, Virginia. We were waiting for orders to become the Flag Ship to the 6th Fleet in the Mediterranean when our captain decided to take us on a shakedown cruise.

The plan was to go from Norfolk to Mobile, Alabama and back. We were making a good-will visit by showing up at their small Mardi Gras celebration.

This trip would be about seventeen hundred nautical miles, one way, and would take four days, one way. We would do seventeen knots most of the time because our top speed of twenty-five knots would cause the ship to "shake" because of propeller damage which the Navy had decided not to fix as it would eventually decommission the Springfield anyway.

Our route around Florida would take us through the Bermuda Triangle. The Bermuda Triangle is approximately 500,000 square miles of ocean with the reputation of not giving back the ships it scoops up. Not to worry, right? Today, the Bermuda Triangle is known to play havoc with one's compass, a geographical oddity that can be adjusted for. It also is an area of convergence for major storms, a meteorological oddity for which you can not adjust.

We departed Norfolk on a clear, sunny day in February. All was well. However, one or two days out, we were hit by a massive "Nor'easter." During a lull, I unbolted my printshop hatch (*door* to you landlubbers) located on the main deck. I saw what looked to be a thirty foot swell headed our way. Fortunately, my hatch was watertight. We were one or two days riding out the storm. So much for "on schedule."

Once the ship was no longer listing fifteen to twenty degrees, I was able to walk about without bracing against the bulkhead or grabbing at anything that was not riveted down. It was now safe to go to chow and to use the "head" (Navy term for bathroom). The latter you avoided at all cost during a storm because of the smell. Nine-hundred enlisted seamen can do a number on the ship's air circulation during a storm.

When circulating around Florida, in route, a Shipfitter Petty Officer told me that they found eighteen split eight to ten inch I-beams. Each split would have to be welded. The splits ran horizontally with the upper section separating from the bottom section and rejoining the bottom with each sway.

That explains the handwritten sign posted to an I-beam outside the library. It read, "Do not put your finger in this crack!" The crack was about a quarter inch maximum, opening as the two halves moved from starboard to port.

I did not hear of any other damage and I was able to complete the shakedown without being bruised and with all of my fingers intact. I thought "Welcome to the Navy, swabbie." Nothing like this ever happened in my home town of East Rochester, NY.

By the way, I was able to go ashore in Mobile to enjoy their mini Mardi Gras. I took in the parade and its' floats and the bands and music and returned to my ship.

Second Amendment Adventures-2

Henry Vth Addresses the Troops. Chapter 2

By Vaughn Stelzenmuller

WELL, NOT EXACTLY. And not George Washington at Valley Forge either. In fact, I knew then it was *not* the best 120 seconds to address what looked like unguided and hyper-restless "troops" on the park field right now.

This is when Johnson shone, looking back on it. One can see the man warming up to a politician's natural environment second by second. In other words, in front of an audience. Which is exactly what the God of Clumsy Situations demands right now, right at this clumsy moment of young politician, long-retired Army officer, and crowd of god-knows-what dumped haphazardly onto the same uneasy parade ground.

At least they aren't pointing their firearms in nine hundred haphazardly, unintentionally menacing directions. Most, anyway.

So Councilman Johnson snaps his fingers off to the left, and a decent battery powered portable speaker shows up, handed to him from one of his acolytes. After a politically-appropriate squeaking from the hand-held speaker, and a few important coughs from the councilman, he leads off:

"Good morning fellow citizen-soldiers, thank you all for coming here today in answer to this important call from your City, and your State. The appreciation....(etc.)."

A great start. It looks like this field of random and disheveled armed clothes horses are beginning to face front, pay attention, and fiddle with their assorted firearms a little less.

"(blah, blah, etc.)"...the work ahead of us, yes, the work ahead of *us* may begin. Those depending on us—you—will rest easier because you answered the call. Shortly we will begin work, I promise, and....(etc.)." continues the good councilman, doing a job I wouldn't know how to start: rousing, but *not* military yet.

"(blah, blah, etc.) ....Captain [redacted] will shortly be coming around to inspect and set up the organization for us. His experience reaches far back to...."

Goddamm liar. Thank goodness.

"(etc.)"....and we do not expect to hold any of you good citizens any longer than necessary."

Liar, but everyone on this parade ground knows that, and it doesn't matter, and HE *knows that. This guy's a genius.*

"So let me let Captain [redacted] tell us how we will begin."

Unbelievable. There is applause, scattered, but applause nevertheless for this basically un-applaudable near-catastrophe! Johnson gets a double gold star from me, and the God of Clumsy Situations just started sweating over the growing sense of actual purpose sprouting on the parade ground.

I do NOT let a loud, unmilitary *squeak* peep from the bullhorn, and then I start off: "Thank you Councilman Johnson for your gracious intro, and your service to City and State in leading this important program," begins a confident-sounding imitation Robert Redford speaking though my own throat viscera.

"We will begin organizing shortly, but first we have to make sure no one gets carried off in that ambulance to your rear from gunshot wounds. And we only have one ambulance." A general rolling chuckle wafts around the field, heads bending to the back of the field, and general nudging and shuffling, likely because everyone "in the know" KNOWS that NO ONE assembled there could possibly be so stupid as to cause a random gunshot wound. At least, everyone either knows, or is too intimidated to fiddle with their firearms. For a few lucky minutes, anyway.

"Now, please lay your firearms down in front of you, safeties ON, and pointed to the right," continues geriatric, but more severe Robert Redford, ever-so-slightly more edge in the diction, "and do so carefully, and NOW!"

Miracles continue. No gunfire echoes through the park from some dumbass with clumsy index finger. Or foot. Or hell, nose.

A few minutes more of calming paternal Robert Redford instruction will suffice to put everybody in the right frame of mind: i.e., "*don't screw up until the old man passes by*." That's me. That's the idea, no? Sometimes it actually works, and if it doesn't, then just pick a different actor to mime. Time for the *old man* to tour the random military shaped mass assembled in this unsettled place.

****

Time for *David Niven* to take over and make the rounds with a now-confident Councilman Richard "Jake" Johnson at the side. At this point we need to carefully choose volunteer subordinate leaders for this latent mob ("You, you, and you").

Uh, oh! Dare my eyes see random arms, elbows, and fingers reaching toward their recently grounded firearms?

[T.B. CONTINUED: *Chapter 3. David Niven Forms an Army*, p.169]

ENLISTED 2, OFFICER 1

Humor of an Army Mortuary Identification Specialist

By Gary Redlinski

I WAS ASSIGNED TO THE ARMY MORTUARY on Tan Son Nhut Airbase, RVN. I spent 24 months (68-70) there. Although we approached the task of identifying our deceased with dignity and respect, we also found moments of levity to help us through the process.

Only those who have experienced this solemn responsibility can genuinely comprehend its magnitude. Warfare possesses an unyielding manner of shortening existence beyond what it already is.

Sometimes our assignment had it's rewards, even if some was pity from the rest of the military....

When in our "off" hours, partying off base **and** after curfew, was not out of the question. Our Military Police friends, residing in barracks across from our barracks would round us up and transport us back to the airbase, with us hooting and hollering in the back of their Jeep. As we got closer to the main gate of Tan Son Nhut, they'd stop and tell us to try to maintain our composure and **shut up** until we passed through the security guards at the entrance. They would tell the guard that they caught us off base after curfew, and that they were taking us to our company CO for discipline. It's a good thing we knew them and they knew where we worked. I think they felt sorry for us and what we had to do.

Stress relief can come in strange places and at strange times....

At the mortuary, we had 5 walk-in coolers to hold the remains until they were embalmed and ready to be transferred home. Where we worked there was no air conditioning. Only overhead exhaust fans to keep the air fresh. Being in a metal building with no insulation, it could get pretty hot in that tropical climate. Occasionally to off set the heat we would take a break and spend some time in the coolers having a cold beverage, (which we kept there for just for that reason.) Some may think it weird, but we found comfort in going in there and cooling off among our fallen brothers and telling some jokes, with our fallen comrades ("listening in") relaxing together.

Sometimes you shouldn't mess with the enlisted....

I remember a time, during my last tour, while working on a remains, a colonel (who wasn't authorized to be in the ID lab and embalming area) came in as if he was going to get his way and demanded to know where his fallen friend, a Major, was to get his personal effects (a watch and ring). I was the first person he encountered. After addressing me in a demeaning manner, because I was only an E-5. I calmly tried to explain to him that there were no personal effects on the remains, that they were removed before they got to the mortuary. He didn't believe me and demanded to go find him and check it out personally. There are times when rank doesn't always have it's privileges.... As he really started annoying me. I said to him, "Well if you really insist, I'll tell you where he is." The other guy working with me, knew me well enough, that I'd probably do something a little crazy. He looked away so he wouldn't be seen snickering. I told the colonel to look at the 4th embalming table where one of our embalmers was working on a remains. (I still don't know, to this day, if that was the officer he was looking for). Then (I told him), if he wanted, go there and look for himself. All of a sudden he quieted down and started looking a bit pale. Then he demanded to see our Officer in Charge. So I calmly told him where he could find him. After a while, I saw the officer being driven away by his driver. Shortly after that, our commanding officer came in looking for me. He asked me what the h*ll did I say to him? After I told him the story of what had transpired, We all started laughing, and I was told I should be more cordial the next time.

At the Stroke of God-O'clock

By Vaughn Stelzenmuller

YOU SEE, GOD (HE AND/OR SHE) can make "rules" to guide the military better than any government. It doesn't matter if this *STRAC* ("describes a well turned-out Army soldier") new lieutenant is the handsome model soldier involved or not. Fresh from officer basic training, then Ranger training, then civil affairs school, here stood I, ready to impress even George C. Patton his-own-self. I wonder if Big He looked down on me with doubtful eyes as I prepared for a seemingly routine field training exercise. Map reading, to be exact. General Patton's spirit may have only snorted.

One humiliating night in 1968 at Ft Hood, Texas, found me leading a platoon to a distant map reading objective or two, and then supposedly back to base camp. One of the thirty of us found, after two and one-half hours of slogging in the *pitchest* dark I have ever seen, that *one* of us was lost. It is not good for the Leader to be the lost sheep in the bunch. I could feel Big She whisper (stereo!), "Have you found yourself yet? Lieutenant?"

"*Shit, shit, shit*!" said the lieutenant's fast-draining career to itself.

As Staff Sergeant Gordon and I sat down on a log to "figure things out," Sergeant Gordon put a reassuring hand on my knee, pointed to the sky and said, "There's the North Star. Base camp is a few degrees direct south azimuth, so maybe we should head back in opposite direction."

"Man, am *I* glad *I* thought of that!" grumbled I to no one else but my own sorry self. No answers, anyway. It was God-O'clock in the morning, and I was truly thankful for Sergeant Gordon. An hour later, we only overshot the base camp by a few hundred meters.

God must love and trust enlisted men, and that's why He made so many of them. So She sprinkled in a few junior officers for bad examples, just to prove it.

No PhD Needed

By Cindy Blair

I REMEMBER THE PINNACLE OF MY CAREER in the US Navy. I was working in an office doing research on hurricane computer models and global warming. I could savor the images of tropical weather patterns and pictures of ice melt from the satellites. In the afternoons the research team would meet to discuss the day's findings. Three civilians with PhD's and I were on the team. I felt way out of my league but they seemed to value my input.

One instance where I felt important was when a U.S. weather satellite fell out of orbit, as it was expected to do, but the new one hadn't been launched yet. Normally, the data from the satellite was gathered and processed every 3 hours by the mega computers to put out forecasts. When the U.S. satellite went out of commission they relied on data from British satellites and that only processed every 12 hours! Computer forecasts became sketchy and inaccurate. After about a month, I learned that Britain didn't have its own satellite. "Where are they getting their data from?" I asked myself. Using a high-level security on the government computer I was able to look at the source of the weather data being used by Britain. It turned out that they were receiving their data from one of our Asian allies. I brought an idea up the chain of command that since our Asian allies were providing the data to Britain, we could ask them to provide us with the much-needed data also. And they did!

Another important contribution I made was when the computer models were recording sea surface temperatures in error in the Great Lakes. The buoy thermometers were recording 38 degrees to 42 degrees Fahrenheit. The model kept changing them to 32 degrees Fahrenheit. I asked about this anomaly in our afternoon briefing. One of the PhD's brushed me aside and said, "We've noticed it but it isn't high-priority; just keep an eye on it."

A couple of weeks later I queried again about the water temps being reset by the model to freezing. That same PhD who had brushed me aside lit up his eyes, became very excited and said, "The model algorithm thinks it's ice!" Just by saying the word, *freezing*, it triggered the answer. Yes, those were my glory days; too bad they couldn't last.

Previously published in "Afterwords," an anthology from the Veterans Writing Workshop 2020 Hudson, NY

Officers

By John Steele

THE WRITING GROUP HAD A PROMPT that dealt with the interactions between enlisted personnel and officers. But I don't think I can follow the prompt to a T. In the prompt, I am told that while I talk about officers, I must do it "while maintaining a respectful and lighthearted tone throughout." I don't think I can do that. No one can expect an enlisted person to write about officers in a respectful and lighthearted tone throughout.

Early on in my Army career, I had very little interaction with officers. Much of my early career in the Army was spent in line units, and I was at the bottom of the food chain. As a private or a specialist in a line unit, I dealt with my immediate supervisors, who were an E-5 Sergeant Team Leader, or an E-6 Staff Sergeant Squad Leader. Going up the chain from there, next there was the E-7 Sergeant First Class Platoon Sergeant, and an E-8 First Sergeant. As long as you didn't screw up royally, you never really interacted with an officer. Depending on your level of screw up, it would be handled by the Squad Leader, or the Platoon Sergeant, or heaven forbid the First Sergeant. Anything above that, you were really in trouble and you were probably being brought up on charges under the Uniform Code of Military Justice.

Did we have officers speaking during company formations? Yes, of course. But you learn very early on that you don't really have to pay any attention to them. The higher the rank of the officer, the more they just love to hear themselves talk. They could go on and on, blah blah blah blah blah. We would just tune them right out. Much of it didn't pertain to us anyways. Anything that was really important, our squad leader was going to reiterate to us in a much briefer and more succinct way afterwards.

The other issue is there simply weren't many officers in a line unit. My first company had about one hundred forty soldiers in it. If fully staffed, there was a company commander, the executive officer, and three platoon leaders. So there were five officers for my company of a hundred and forty.

When I became an instructor in the reserves, there was another issue with officers that was unique to the Rochester area. My unit was in Webster, NY, but we were part of the 98th Division, which was located on Goodman Street in Rochester, less than ten miles away. Being a division headquarters, there were tons of officer positions. Underneath Division, there was Brigade. My brigade, 7th Brigade, was also located on Goodman Street. Luckily, my Battalion was located in Schenectady, two hundred miles away from us. But Battalion liked to come visit us sometimes, maybe so they could also swing over and suck up to Brigade and Division. And they used to get paid mileage. They would put themselves on TDY (temporary duty) orders, and get paid something like thirty cents a mile plus a per diem for each day. Thus they could get paid a couple hundred dollars plus hotel in addition to their regular weekend pay. It was beneficial for members of Battalion financially to visit us whenever they could think up an excuse to do so.

So while my unit had a Commander, and an Executive Officer position that was empty more often than not, we still had officers visiting from Battalion, Brigade, and sometimes Division. Most of the time, they were on what they called "morale visits," where they would show up for a half hour or so to observe, as they ate donuts and drank coffee. Then they would tell us what a great job we were doing, but then tell us we had to do this and this and this differently. They knew, for they observed us for a few minutes from a distance while they greeted each other and ate our donuts and drank our coffee. It didn't matter what we were doing, nor whether we were correct or not. As officers, they inherently had to correct us, even when they were clueless on the subject matter. But then they would leave. While they were there, we'd be polite and courteous, nod a lot, and just wait for them to go away.

There was one other time where we had to interact with Battalion. Once a year we had a command inspection. A command inspection is when Battalion comes down and basically tells you everything you've done wrong. As the expression goes, shit runs downhill. Training command would yell at Division, Division would yell at Brigade, Brigade would yell at Battalion, and Battalion would yell at us. There were all sorts of mandatory training and tasks that needed to be done periodically. The goal was always 100% compliance. It was rarely met, but that is what you strived for. There were numerous times where Division, Brigade, and Battalion numbers were all under 50% for various tasks. So Battalion would scrutinize our numbers, and question why were we not at 100% compliance. Our numbers were almost always way above Battalion and Division. They would bring up some mandatory class, and we would be at something like 94% compliance, and of the three soldiers who hadn't done the training, one was at an NCOES (non-commissioned officer education system) school, one recently had surgery and had been out the past few months, and one was being processed out for not showing up for drills. We submitted his packet for unsatisfactory participation five months ago up to Battalion, and we didn't know why he hasn't come off our books yet. We were waiting to hear a status update from Battalion. Yes, we would say this to Battalion. Yet we would get dinged for only being at 94% even though Battalion, who was inspecting us, was hovering around 45%. It was hard to take them too seriously.

Did I ever interact with good officers? Yes. Many of them were good. Did I have poor ones too? Absolutely. We had one major as a commander, and we would consistently have our 7:00AM morning formation without him, as he rarely made it to the unit on a drill weekend before 7:30 or 8:00. Good things about him: he would sign anything we put in front of him, and most days that we had drill, he would want our day to be over and for us to be out of there and be heading home by 3:00 or 3:30 instead of the usual 5:00 or 5:30 or 6:00.

The worst thing he ever did? We were invited to be in the town of Webster Memorial Day Parade. We had some cool Army vehicles, which we drove in the morning parade. Army was paying us for the entire day. It was a beautiful day, sun shining. The parade went from the high school to the cemetery in town where they had a ceremony laying wreaths for service members who died in various wars. As we were approaching the cemetery, the commander got on the radio and said it was too nice of a day, we weren't going to waste our time sitting through some boring ceremony. I was driving the first vehicle, and when we came to the entrance to the cemetery, I pulled in like the rest of the parade. The commander was in the second vehicle, and he ordered the driver to not pull in, go straight. They were going back to the center. And he ensured all the other vehicles followed his vehicle. It was beyond embarrassing. What a disrespectful display, on Memorial Day, by a Major, in uniform. He didn't want to honor fallen brothers in arms because it was too nice of a day and he had a picnic to go to later that afternoon.

I had the keys to the motor pool that day, so they couldn't park the vehicles when they got back, they had to wait anyways. The parade was about an hour. The ceremony was maybe an hour. It was a two hour work day, we were getting paid for the entire day, and he still wanted to cut it short. When I returned, my commander was pissed and said he was going to write me up. I disobeyed his order, and I turned off the radio as he continuously called for me during the ceremony telling me to get back to the unit so everyone could go home. I begged him to write me up, because I was definitely going to inform our higher headquarters of the circumstances. Even with that event, a few months later Battalion gave him a glowing Officer Evaluation Report and he was moved into some staff position.

I have one other officer I must talk about. Our unit were Combat Engineer instructors. Our annual training was spent down at Fort

Leonard Wood working with regular Army personnel instructing soldiers going through Advanced Individual Training to become combat engineers. Combat engineers must be male. It is one of the few MOS's (Military Occupational Specialty) that still prohibits females. But our Battalion and Brigade was made up of various other MOS instructor units and drill sergeant units. So our Brigade commander happened to be female. When she arrived at Fort Leonard Wood, she had flown into Saint Louis, and the airlines lost her bags. So she only had the clothes that she flew in on. The best way to describe it is she looked like she was wearing a slutty prom dress, complete with painted on make-up and heels. Yet she insisted on visiting all of the various ranges that we were working on. So there she is, showing up at the demolition range, and the Bailey Bridge site, wearing a too tight, and way too revealing dress, with five or six inch stiletto heels. We had to explain to our active duty counterparts that she was in fact our full-bird colonel brigade commander, not some high-end prostitute. She did her best to ruin our credibility.

Slow Night at the Mortuary

By Gary Redlinski

ON OCCASION, OUR NIGHT SHIFTS at the US Army Mortuary on Tan Son Nhut Airbase, R. V. N. would be slow and uneventful. To combat the monotony, a group of four or five of us would gather in the XO's office to play cards and pass the time. These games could last for hours, with occasional breaks for snacks or conversation.

It's important to note that at the mortuary, our only comrades were those who had fallen in service. There was no hierarchy based on rank - everyone was equal in death. This meant that while we respected each other's roles and responsibilities, there was no sense of superiority or inferiority based on position.

During one particularly long card game, Lt. M. was having a string of bad luck and losing hand after hand. We (the enlisted soldiers) couldn't help but tease him a bit about his misfortune. Suddenly, Lt. M. jumped up from his seat and grabbed a yellow Smoke Grenade off the top of a filing cabinet.

At first we thought he was joking around, but as he pulled out the pin and held onto it, we realized he was serious. As he struggled to get the pin back in place, we started to panic - what if this wasn't just a prank? Eventually Lt. M. gave up on reinserting the pin and instructed one of us to open the window so he could toss the grenade outside.

Unfortunately for us, our location near the base perimeter meant that there were guards stationed nearby who quickly noticed something amiss. Soon enough, several Air Force Security Police vehicles arrived at our doorstep.

Lt. M. 's explanation for what happened was flimsy at best - he claimed it was all part of a "training session" for new troops and that everything had been under control until he accidentally pulled too hard on the pin. The rest of us tried not to laugh as we watched him squirm under questioning from security personnel.

Once they left, however, it was impossible not to tease Lt. M about his ill-fated attempt at entertainment during an otherwise dull shift. In hindsight, it's clear that what happened could have been extremely dangerous - but at the time it felt like just another way to pass the time when nothing else seemed interesting enough to do.

General Electric Issues Orders

By Dana Hill

DURING MY TIME SERVING IN THE AIR FORCE, I had the opportunity to work on medical equipment. One day, while I was waiting for a call from the manufacturer of an x-ray machine that I was currently working on, I received a call from one of my young airmen. He informed me that he wasn't feeling well and was calling me from home to let me know. As we were talking on the phone, suddenly a third voice interrupted our conversation. It was the base operator who announced that we needed to terminate our call as a *General* officer needed to speak with us.

Initially startled by this unexpected interruption, I couldn't help but wonder why a general officer would want to speak with me. However, the base operator's tone was firm and she repeated her request for us to end our conversation. We quickly hung up and almost immediately after, the manufacturer I had been waiting for called me back. Curious about what had just happened, I asked if by any chance they had gone through the base operator.

To my surprise, they confirmed that it was indeed General Electric (Co.) who had been trying to reach me and that I needed to speak with them urgently. Despite being caught off guard by this turn of events, I couldn't help but appreciate how helpful and kind the base operator had been during our brief encounter. Her actions demonstrated how important it is for each member of the military community to support one another and ensure that communication channels remain open at all times.

It's amazing how unexpected events can occur in life—even during seemingly routine moments like repairing medical equipment. Nonetheless, this experience taught me to always be prepared for anything and everything!

On Parade

By Vaughn Stelzenmuller

PICTURE THIS NEAR-PERFECT ARMY CAPTAIN readying to leave for a special joint U.S.-German formation, 1970s Stuttgart. The morning dawned near perfect also, sunny and warm and announcing a splendid German summer day ahead. A perfect day for this happy event.

Parades and formations aren't always greeted with either the usual tedium or irritation in Army life, these just *are*. Today, though, I felt the rare happiness of looking the picture of one of those soldier color plates found in those glossy Osprey military paperbacks. My mirror told me I was right! Pressed Army class-A "greens," decorations perfectly in place (no ostentatious medals, please!), shoes a high polished black, all hanging sleeve threads cauterized before leaving home, hat brushed. What a perfect Army captain stared back at me from that mirror.

Joining me to leave our workplace for the formal morning formation, my deputy, Staff Sergeant Locke sported his usual professional stern-humorous face, and we started out the front door.

As we emerged into the bright sunshine from our work building Sergeant Locke leaned professionally forward toward me and whispered the most important sentence of the morning:

"Sir, you're advertising."

Holy Toledo and General Arthur Lord Wellington! I didn't have to look down to know that my fly was unzipped! Any skilled soldier knows how to keep walking, pretend to bend over and fish for something in the trouser pocket, and zip up without cutting stride. Otherwise, that near-perfect day, I would have indeed been world-wide, *on parade*.

DON'T SCREW THIS ONE UP

The Shackleton, the Cockney, and the Yank

By Charles F. Willard

WHEN WE LANDED AT CHATEAUROUX AIR BASE in central France late in the afternoon of 3 July 1965, our Air Force C-130 crew had been flying Cold War redeployment missions around Europe and the Near East for days with minimum rest. Finally, we were due for a break—forty-eight hours on the ground, uninterrupted by flying.

Tomorrow would be the 4th of July. At that time, in the first half of the 1960s, the trauma of World War II was still fresh in European minds. Tomorrow there would be a commemoration of America's part in securing that victory. Our NATO allies had assembled a menagerie of planes and stunt teams for the holiday air show, and we could watch it. From the ground!

The "Follow Me" truck led us to our spot on the crowded ramp, where a wand-waving ground crewman motioned us into our parking position next to a strange-looking aircraft with Royal Air Force insignia.

It squatted there on the tarmac like an ugly gray toad.

I recognized the incongruous-looking plane. It was the Avro Shackleton, named for Ernest Shackleton, the famous British Antarctic explorer. It also had an explorer's mission—to go on the hunt for enemy submarines.

Lugging our B-4 bags, our tired gaggle stumbled aboard the waiting crew bus and headed for quarters. We got cleaned up, then, tired but thirsty and famished, made a beeline for the club.

The Officers' Club was mobbed with an international mix—French, English, Italians, Turks, Norwegians, Canadians, Germans, Greeks—most of them here to fly and show off their aviating skills in tomorrow's big show. It was a loud and boisterous atmosphere, everyone competing, telling their own war stories—"There I was..."—in several different languages.

Mingling in the crowd, I found myself standing next to a Brit.

"Oi sigh, Mite, did yer just come in on the 'erc?" He used the nickname for our airplane, the C-130 Hercules.

"Yep. We've been buzzing around Europe and the Middle East, repositioning troops and supplies for the last month or so."

"Saw yer land. Oi envy yer, Mite. Yer gets to floy all over. Oi floy that "Shack" you parked next ter. We don't get ter see much. Just orbit over the ocean lookin' for Ruskie pig bowts, yer' know? Oi'm Floight Leftenant Jack 'arrison."

"Lieutenant Chuck Willard. Pleased to meet you, Jack." We shook hands. He pumped mine as though we were long-lost friends.

"How about a beer? I guess you Brits call it 'a pint'."

"Roighty ho. Oi see Oi've got me a new best bud, Oi do."

I wondered if my American accent delighted him as much as his Cockney one charmed me. We hit it off right away, swapped stories, and ended up eating dinner together, still gabbing away about our flying experiences. During the evening's conversation, it developed that Jack had a problem. He was supposed to fly his "Shack" in tomorrow's air show—a low-level demonstration—but the Flight Surgeon had grounded his copilot. Ear infection.

"Bloimey. Bugger awll. Blarst me luck."

"Gee, that's too bad, Jack. I'm sure you were looking forward to it."

"Yez, Oi surely was.... Sigh, Mite, Oi've got a bit of an idear 'ere. *Yew* could be me blarsted copoilot!"

"N-N-Now wait a minute, Jack. Is that even legal? An American copilot on an RAF airplane?"

"Shoore, it tis. Me mites won't moind. Besoides. Ooo's ta knaow?"

"Yeah. But I'm not checked out in a Shackleton. I don't even know what manifold pressure is. The Herk engines are turbos, not 'recips.'"

"Aa, Me engineer 'andles awll that kit. Oi'll do the floyin'. All yer 'as ter do is 'andle the checklists and the rydios and enjoy the roide."

So we sealed the deal with one more pint for a nightcap.

And that was how I ended up with one hour and thirty-five minutes of RAF flying time in my USAF pilot's logbook...." *In me U-SAF poilot's log boook.*"

Bake Sale Saturday

By Vaughn Stelzenmuller

EVERY NOW AND THEN (and then some), we Service people take one more beverage than we should have on a Friday night. Back when our garrison assignment in early 1970s Germany meant working half-days on Saturday, that one-too-many Friday made Saturday painful.

Yes, this particular next-Saturday morning showed me at my desk, trying hard to concentrate at Army-desk kind of garrison assignments, involving a 2-1/2 inch stack of folders as I recall. Phones jangled with that charming 1970s quality of government-issue corded phones, and the smell of numerous cups of coffee from around the building pushed gladiator spears into my troubled head and gut regions.

Staff Sergeant Locke spotted my troubles with a subtle chuckle, but was discrete enough not to comment much aloud. Sometime around ten o'clock a. m. Sergeant Locke jumped up from his desk and headed for the building front door. His desk faced the window to outside so he could see all incoming foot traffic. I heard him calling out, "In here, bring 'em in here!"

Who showed up through our office door but a very cheerful four-wife group with a huge array of cheerful, colorful baked goods! I do remember shouting out "Sergeant Locke!!" and then burying my aching head in my forearms. I tried hard to control my provoked stomach further. These four women were so happy, so pleased they had created this landscape of muffins, shortening'ed cookies, gummy candies, and cake slices of many colors. A loud, colorful, and hugely pleased group.

[A stray thought: I challenge anyone to gag down a small bowl of Froot Loops™ with a hangover.]

After saying all the appropriate and cordial things that should be said, I bought several items from these very happy and dedicated ladies, these good people. They were very pleased. Sergeant Locke was the most pleased person in the building. A better choice of Friday night drink clearly was a good "lesson learned," as we so irritatingly say in the military. It's also good to have a deputy who is deft at disciplining his superior.

Driving in Florida

By Lee Nelson

THE STANDARD IS ASSHOLE.
For all the peril of every
municipality,
this you need know now.
You need Navy Seal A-game,
and A is for asshole.

The Autobahn is the only faster road of lunatics,
but they're Europeans.
They have decency you don't find in Florida.
They use lanes correctly, and stay in their lanes
when 80 is in the way, and any whoops is
manslaughter.

And the old are most everyone,
and they'll kill you just to be rid of you
interrupting their best years in perfect weather.

Then there are the criminals in big rusty bullets full of guns,
and you and me who seek safe passage
across the hottest asphalt that will ever
deliver Disney.

Do or die is implied, and there are no better words.
Even the dead alligators look foolish and innocent.

MILES Gear

By John Steele

IN THE LATE 1980'S, I went to phase one of BNCOC (Basic Non-Commissioned Officer's Course) and we got to use the MILES (Multiple Integrated Laser Engagement System) gear. It was the first time I had ever used it.

What is MILES gear? The overly simplistic answer is it is the Army's version of laser tag. Everyone has a device mounted on the barrel of their M-16 that emits a laser every time a blank is fired from their weapon. Everyone also has a special LBE (Load Bearing Equipment) or vest that they wear, which is covered with sensors, as well as sensors on their helmet. If someone fires a blank from their M-16, it sends off the laser, and if that laser hits someone else's sensors, a very loud tone will sound on the hit person's gear indicating they are "dead." It can also sound off a quick "beep beep" indicating a near hit.

They also have MILES gear for vehicles, aircraft, and all sorts of other weapons and weapon systems. We were working on squad level tactics in Phase I BNCOC, so we only had the equipment on our helmets, LBE's, and M-16's.

The way it was supposed to work is there is a key on the laser device on your weapon. If you get shot indicating you were dead, the loud tone goes off, and the only way to turn off the tone is to remove the key from your weapon, and insert it into your gear to silence the tone. With the key removed from your weapon, your weapon will no longer work. Remove the key from your LBE after you have been "killed," and the tone will again blare. Instructors have the "god" key which will reset your gear, turn off your tone, and bring you back to life, allowing you to reinsert your key into your weapon.

That is how it was supposed to work. How did it work? Many soldiers from BNCOC classes prior to ours kept losing their keys, so the staff at our school set the devises that emit the laser to a setting that was always on, and removed the keys. So if you were killed, your tone loudly went off, and there was no way to turn if off until one of the cadre came by with the "god" key. At the same time, there was nothing preventing you from continuing to shoot at other people, because cadre had set your laser in the permanently "on" position.

Now let me tell you about these Phase I BNCOC classes. There were two classes being conducted simultaneously, each with twenty-four soldiers. The cadre decided that they would combine the two classes for this exercise. One class would be conducting squad tactics and movements, and the other class would be the OPFOR (Opposing Forces). Then the classes would switch roles. Soldiers who attended the classes came from several different states, so it was rare if you knew any of the other soldiers before the class. However, after two weeks of these classes, you got to know most of the other soldiers and made a few friends, and maybe one or two others who were not so friendly.

There was one other issue which hindered this exercise. There were very few cadre. There were only two per class, or four, to babysit forty-eight soldiers. As the gear was being issued out of the back of a truck, each person had to be checked to make sure the equipment was working properly. Soldiers had to pair off, and with a cadre, you had to shoot your battle buddy in the head to make sure that your laser worked and their helmet sensors worked. Then the cadre would use the god key to turn off your battle buddy's obnoxious tone. Then you shot him again on his vest to make sure that was working. Then he got to shoot you in the head and in the vest. Everything ran on batteries. They were supposed to be long-lasting batteries, but many were dead. And some of the equipment simply did not work, regardless of how fresh the batteries were. Then you had to figure out what wasn't working, the laser attached to the weapon, or the sensors, or the speaker that emits the tone. So that took time, and people had to turn in equipment that didn't work and get replacements. Cadre was also issuing the equipment. After your equipment was squared away, you had to wait at the edge of the training area for the exercise to begin. It left all the other soldiers with a lot of idle time, and little supervision. What is the worst that could happen?

Someone accidentally shot off a blank. Was it on purpose? Were they firing at someone? Who knows. No one's sensors made any noise. But someone claimed that another soldier was firing at them. So, in response, a few minutes later, the person who claimed that they were being fired at, fired at the soldier who they felt was targeting them. Again, no sensors went off after the blank rang out. The cadre are now getting annoyed, and yelling for everyone to make sure their weapons are on safe. A few more minutes pass, and the soldier who was accused of firing the first shot, and was probably the target of the second shot, definitely fired at the other soldier. It was a kill shot, and the other soldier's gear started bellowing. The soldier who had been killed had to go to the cadre with the god key to turn off their alarm. The cadre is now getting mad, but no one is fessing up as to who fired, and only the killed soldier is accusing anyone, and that soldier has no proof. The soldier who was killed now wants revenge. At this point, there are probably at least thirty soldiers hanging out at the edge of the woods, and all the cadre are trying to issue gear and make sure it is operational. Only a couple of minutes later, another shot. The accused shooter's gear does the double beep, a near miss. The cadre starts screaming at everyone to cut it out, but another shot rings out. Then two more, and finally the accused soldier's signal indicates a kill shot as his vest starts buzzing. The soldier seeking revenge couldn't hit the lake if they were standing on the pier, but their friends joined in to help take out the original accused shooter. The original accused shooter's alarm is now going off, but he doesn't care. He fires a few shots back, setting off more alarms. Next thing you know, one small group has open fired on another small group, and that group fired back. Someone randomly fired at some of the soldiers who were not involved at all, and they started returning fire. Soon I'm guessing at least one hundred reports were heard echoing through the woods, and at least ten soldiers now had their earsplitting alarms going off. The cadre were running around screaming, losing their minds. Total chaos.

How do you punish misbehaving kindergartners, or junior NCO's going through BNCOC? You take their toys away. The cadre decided that first, all the weapons had to be turned in, and only then were all the clamorous alarms turned off of all the "dead" people. The cadre then finished issuing and checking the equipment of the remaining soldiers, confiscating the weapons as soon as they proved operational, until everyone had their gear. Only after everyone was issued proven working gear, and under strict supervision and surveillance, did the weapons get re-issued and we were marched to the edge of the training area to finally begin our training exercise.

Private Skibootch

The following three Private Skibootch *stories are xcerpted from "O'Book," Robert M. Whelan, 2020*

By Bob Whelan

LIFE IS A BUNCH OF OPENING AND CLOSING DOORS. You open one go through a room and out a door on the other side. Going out that door puts you in the next room, on and on.

Skibootch's High School agony came to a halt at precisely the same time he reached the draft age of 18. In WWII everyone was legally subject to the draft, Few exceptions. The controllers of that good news was the Army itself. Skibootch thought he was going to accept an offer from them that would turn bullshit into electricity. No lowdown, unpleasant, experiences for him. Not going to be like High School this time. He was going to College on the Army's dime. He was being offered an opportunity that seemed too good to be true. He was going out of the High School door and expecting to enter the College door.

The Army in 1943 went around the country to selected High Schools and presented what they proclaimed to be a free College education for anyone interested in becoming a needed specialist for the many complicated jobs necessary for their now greatly expanded technical services. They had a name for this emergency effort. It was called the Army Specialized Training Program, ASTP for short.

Such a deal! Who wouldn't jump at one like that? You know who, Skibootch of course. Less than a month after his High School graduation he enlisted in the Army's ASTP Program. It only took a few months for the Army to fudge on its promise and through a series of unsuccessful ways to find an alternative to their original offer, Skibootch wound up, not an ASTP College Graduate Specialist, but as a Private in the 59th Armored Infantry Battalion of the 13th

Armored Division. What often seems to good to be true, often is. Just like High School Skibootch was screwed again. It was beginning to make him angry, and his behavior began to reflect it. He was ready to serve his country and was reconciled to that part of his service commitment, but to the broken promise, no way!

In short order, his behavior turned to finding all the ways to show his appreciation for what he perceived as a screw over. He then decided he would accept Army discipline the same way they fulfilled their promise. There were many stories of how that transpired, but one of the most bizarre was how he achieved the Good Conduct medal. Not everyone did, and after thinking his way through it he came to the conclusion that the Army's vision of good conduct was based on risk taking, not being nice.

Being adept at finding ways to break through the cracks in unwanted domination such as being misled and broken promises he spent the balance of his time in the 13th Armored Division training camp, breaking as may rules as one could imagine.

On the surface his behavior would be considered exactly the opposite of what a training camp soldier should be. He was at the low end of the model soldier model. One of the most telling examples happened after he returned to camp after spending the night in the Fort Worth, Texas jail for being involved in a barroom brawl. On reflection, it would be reasonable to assume that those in charge of awarding Good Conduct medals would be aware of any occasion involving police detention, especially when the soldier was missing from roll call, and thus being AWOL. That assumption being the case, it has a lot to say about what the Army considers good conduct. For an Infantryman fighting and risk taking is his number one asset.

On a hot 100 degree overbearingly heated day at exactly noon Private Skibootch arrived at the A Company Parade ground, after being released from his jail detention. He was all bloodied and disheveled, but he got there just in time to be awarded, in front of his less favored comrades, the Good Conduct Medal.

Private Skibootch would soon be on his way to where the real fighting was going on and had already demonstrated the way the Army rewards risky behavior. They liked fighters. Private Skibootch was thus rewarded and on his way to becoming a Private First Class

Private First Class Skibootch

By Bob Whelan

CAMP DUTY AND COMBAT DUTY are two different things. Camp duty has a lot of unpleasant bullying by the upper ranked. The game you play there is to find ways to return some of the unpleasantness. A risky game to be sure, because if you get discovered even worse punishment occurs. Combat duty is the same, except that the adversary is your enemy not your bosses. You and your bosses are all on the same side now and everybody's duty is to help and protect each other. In Duty language: "To have your buddies back."

Low rank foot soldiers, like Infantrymen are mostly rotated as "Point Men." Point men's duties are to scout where the enemy is and direct fire accordingly. It is a high-risk job, and the survival rate is low and mostly miraculous. But risky behavior it is, and someone has to do it. It has the same degree of duty as does existence in Coved 19 protection. Terminate the enemy before he terminates you and your buddies. You keep him from terminating you. You do everything to avoid that risk. You obey orders. You practice protecting those who protect you. Everyone on your life is a part of this protection. All are in it together. Exposing yourself to the enemy is risky. To do it one must be a trusted risk taker.

For all its grievous bullying mannerisms the Army has its own brand of acceptable conduct. It appreciates those who do not shy away from confrontation. You might call them "Risk Takers."

When Skibootch received his Good Conduct Medal after risking punishment from both friend and foe, it was an insightful journey into the true spirit of military duty. The medal seemed to be a reward for demonstrating the quality of risky behavior. Skibootch, believe it or not, not only got the medal, but a promotion as well.

Within a short time of the barroom brawl fiasco, Skibootch was promoted to Private First Class. One could surmise that down deep the Army recognized the value of a fearless risk taker. Risk demonstrated in any form is risk. In an Infantryman it is performance value. In one of his first Combat actions, in true Skibootch fashion screwed up in risky behavior by being where he shouldn't have been, while releasing enslaved prisoners from their long agonies serving their oppressors. In so doing, he became detached from his unit and spent the better part of four days reconnecting. In that time, he and his partner, a Private named Larry, joined another Infantry group and performed point man duties with them.

Shortly after joining that group, he, and Larry, scouted and entered an Artillery Spotters location and took several prisoners, including two officers. More importantly, he acquired from their location a beautiful pair of Afrika Korps Field Glasses that ultimately became a symbol of how he was able to see the world

Afrika Korps Field Glasses were to accompany Skibootch through all of his WWII experiences and right up to about ten years ago when they mysteriously disappeared. They became a symbolic connection to life as Skibootch experienced it.

WWII German Afrika Korps Dienstglas Seeing the world through that Dienstglas was a big part of what made Skibootch, Skibootch

There are many sides to every experience. The story of Skibootch's Good Conduct Medal and his becoming a Private First Class are just the beginning.

They pick up again as Skibootch looks at why he became a Combat Infantryman in the first place. It was all part of how he was able to see what real life and excitement all are about. They focus on that picture, just like the Dientsglas does on its target.

Dienstglas Skibootch

By Bob Whelan

THE WAR IN EUROPE ended in early May 1945. Not Skibootch's luck though. Skibootch could now see things he couldn't before. Hidden obstacles were now becoming visible. Playing the game of life became more strategic and less dependent on chance. Risk taking became more viable and manipulative. What the hell, he was able to see better. That was it.

Skibootch and His Dienstglas at Goering's House, May 1945. There he is, Dienstglas draped over his chest, in front of what's left of Hermann Goering's Mansion at Berchtesgaden a few days after WWII in Europe ended. His 13th Armored Division had advanced all the way to the Austrian border and even captured the town of Braunau where Hitler was born. It was only a short journey after the war to come to Berchtesgaden and witness history.

Hitler's eagle's Nest was only a few hundred yards away and it came through unscathed. Goering's was bombed by the RAF on April 25, 1945. Skibootch and the Dienstglas were on their way to taking a clearer look at life. To understand why risks need to be taken, and why relationships matter. The Dienstglas became the key to this understanding. Everything in that picture of him at Goering's house are part of its meaning.

The house, to start, has a clear message. It metaphorically represents in its demise the result of a life lived in self-service. By taking from others but not serving them in exchange. They had a right to sharing that house as did Goering had an obligation for the same. Those with whom he denied, shared their denial, His house was destroyed. His house metaphorically was the Third Reich which went about denying their neighbors houses. What goes around comes around. His house was denied. The Dienstglas was the vision of that denial. It had the various lenses and instruments to see the relationships involved in that process. The house being discerned up close. Being discerned in relationship with distant houses. Having the tool to correct the efforts to share its nonentity. He who destroys will be destroyed by his destruction, as will his house.

The close-up lens when reversed becomes a faraway lens. The calibration inside helps correct the aim of the force that is aimed at the target. In Goering's case the target was the house he desired. Fancy and exclusive. The house where one lives who has achieved their purpose. Purpose being their way of living. The Dienstglas is a metaphor for envisioning that process. It takes into account the multiple factors that are a necessary part of that achievement.

Dienstglas is metaphoric. Simply put, it is a comparative way of looking at life.

Skibootch is metaphor. We are all metaphor.

* * * *

Stories "Private Skibootch," "Private First Class Skibootch," and "Dienstglas Skibootch" were previously published in *O'Book*, by Robert M. Whelan, ©2020

Second Amendment Adventures-3

David Niven Forms an Army. Chapter 3

By Vaughn Stelzenmuller

WELL, NOT EXACTLY. Not right at this second, anyway. Time for George Washington (again!) and Abraham Lincoln to rise up.

Washington was a genius. One way or another he found the magic working formula for civilian-military control over the armed services, and it works here—the civilian politician (and Constitutional superior!) Johnson with ultimate responsibility, and the military person, me, to pull it all off on the *battlefield*. This is definitely turning into a George-Washington situation: for Washington it was setting up and controlling a wild collection of future "soldiers" for which there were no guidelines, no strategy, and no pay. Well, now *shoulder the backpack* and move on.

What would a lightning-fast home defense look like? Abraham Lincoln had some experience here, I recall.

What if they asked me, and I have stumbled onto, to be the "battalion" commander of such a grouping? I have been asking myself ever since showing up on this overcrowded park (now parade ground). Well, here I am.

"FAH-H-H-H-H-H-L-L IN!!!" I shout through Jake Johnson's microphone, standing like Robert Redford not trying to make a fool of himself. Time for some aside explanation to our great and blessèd non-military readers. David Niven explains it from here....

"A good commander does not bury his or her face in his or her hands when a really bad scene next shows up in front of that same face," the good Mr. Colonel Actor David Niven preaches.

Because what shows up next is half the crowd getting that dopey "HUH?" look on their faces, and another half scratching their individual differently-facing heads about where this fall-in has to happen, and another half trying to remember how to do all that fall-in stuff from military or policing days, and the fourth half making snickering and snorting sounds to cover up their fright, or boredom. In other words, *Americana* all around.

Truth be known, this is now the ripest time to select subordinates. Officers first, then the more crucial *noncommissioned* officers—our indispensable "sergeants."

Fortunately among all those four halves we see a 5% who really do look like they have some mental organization or experience, or both, and truly seeming like they welcome some kind of assignment. Congratulations. We just spotted our officers, and far more importantly, our "sergeants." The difference will be that the supposed-officers have that "I'm entitled to peoples' respect" look about them, and we know they would rather die physically than get shown wrong or cowardly. Perfect officer material.

And the "sergeants"? These good folks probably know they are going to die either from boredom or by some *crazy* in the crowd, but have an unspoken quiet about them that allows them to put the thumb on subordinates to do the right things, and which subordinates believe their own chance of dying is smaller because of his new sergeant. All of the above is good, if good selections can be pulled off, and quickly.

Ah yes, really important: Sergeants are cagy, willing to smack someone upside the head to get the attention of neighboring "strays." Heck, it doesn't matter if the smack-*ees* were paying attention or not. The *non*lethal disciplinary example is worth it. This is the value of American military genius: soldiers like discipline, enjoy showing off that they actually don't, and everybody adjusts to the informal and directed "way." Ninety percent of the group will "get it" and go along, even laughing off some of the theater. That other 10% would get fired from any job anyway. And the smack-*ees* get a great "war story" to tell their grandchildren.

Oh yes, I already had four persons whom I dragged along as assistants earlier (and all right, as "comfort horses"), and who fell in line automatically as my subordinates, who after the next two days will hate my guts from envy or much direct-speak (from me), but who are utterly essential right now. I can't be everywhere at once as needed, so they have to fill in some gaps. They will do an excellent job, get little credit for their results (except a paltry, but excellently well-written commendation from me), work their butts nearly to death, and *also* get a good war story to tell their great-grandchildren. Perfect subordinate officers. Then a loyal, top-notch aide-de-camp can be spotted in the crowd later; look hard for a brainy, handsome butler, and we have it. Get a bad aide-de-camp and the commander is screwed. That's a *military term*, you know.

Now I will have the pleasure touring the assembled recruits to nominate subordinate leaders who will truly orient the group. Or get an estimate of the size of this disaster early on. Let's see what we have here to work with. Councilman Johnson accompanies me, and he will get the credit later for the job David Niven does right now in forming this mass into something directed. If it doesn't work, he may have to try explaining it later. This is how excellent military-civilian relations work. AND, he will help keep me out of jail. Especially if it turns out that no one gets shot....

[TO BE CONCLUDED: CHAPTER 4, *Light at the End of the Gun Barrel*, p. 246]

BETTER LUCKY THAN GOOD

Initial Operating Experience

By John P. Bland

ON A VACATION TRIP TO HONOLULU, my child bride of fifty-plus years and I visited the Hawaii Museum of Flying at the Naval Air Station, Barbers Point. There we saw a US Marine Corps A-4E Skyhawk Bureau Number 151036, which, like me, had survived its military service mostly unscathed. The beautiful old bird, which had been manufactured in California in 1964, was struck from the rolls by the Navy Department in 1992. For a combat jet airplane, 8-10 years old is middle-aged; this airplane's career was 28 years long. The single-seat, single-jet-engine Skyhawk was a bomb-, rocket-, and gun-carrying attack airplane designed to provide close air support under the control of a Forward Air Controller (FAC) to friendly ground troops by dropping its ordnance on enemy troops. When I spotted its Bureau Number, I immediately remembered flying this very jet on my first sortie in Vietnam in 1972.

Flying at 15,000 feet above the ground, I spotted soupy clouds to the far west and a higher, solid cloud layer directly overhead, neither of which concerned me. Immediately below, through holes in a broken layer of low grey clouds, I noted the luxurious green fertile terrain. My worry was that in this hot, rainy monsoonal season this low scud might clot into a solid undercast in our target area, which would require abandoning our mission. My wingman and I were well prepared to accomplish the close air support operation even with opposition from the enemy, but the tropical monsoon was always a threat to flyers.

This sortie was my first combat mission. The squadron's senior leadership, all with combat experience in Vietnam from years before, had led the missions that morning in May 1972, which was our unit's first combat day since our arrival 36 hours before. I was honored to have been chosen as the first of the young guys to lead a day one combat mission. The voice in my head hollered, "Don't let me screw this up." I wanted to do a good job for those in rank above me who apparently had confidence in my abilities as a pilot and a leader and for my unit, too. Marine Attack Squadron-211 (VMA-211) had an enviable reputation, probably because of its three previous tours in Vietnam beginning in 1965.

As we flew north, openings continued to appear in the undercast, and I saw the country, low and flat with a collection of small roads, lazy streams, verdant forests, isolated villages, fruitful rice fields, abandoned rubber plantations, and hundreds of rainwater-filled bomb craters. Attacking our target would increase the number of craters, and I thought, "Am I going to add to the misery of the people below by providing more breeding grounds for untold numbers of malaria-carrying mosquitoes?"

Other concerns flooded my mind. Could my wingman Woody and I expeditiously find the target and the airborne US Air Force FAC? The FAC was the one who would identify the target and report the results of our bombing. We had no sophisticated navigation equipment, only a compass, a paper map, and a rudimentary radio navigation set designed in the 1950s. Would we have enough fuel remaining on board after our mission to make it back to base? What about enemy opposition? We had been briefed before takeoff that the enemy might be using hand-held surface-to-air missiles (SAMs) and light anti-aircraft artillery (AAA). That was why we were flying above the threat at 15,000 feet. As we progressed northward to the target area and contacted our airborne FAC, I had time to think about how I was flying this mission on my squadron's first combat day in Vietnam, halfway around the world from home and family.

Our weapons loadout for that combat mission had been a standard one for the diminutive Skyhawk; one 300-gallon drop tank full of fuel, twelve Mark 82 500-pound bombs and 200 rounds for the dual, internally mounted 20 mm cannons. We stopped in the weapon arming area where ordnance men from our squadron pulled the bomb safety pins and armed the guns. Then we launched. The venerable jet was loaded to its maximum takeoff weight as it lumbered into the hot, moist air after a long takeoff run.

Airborne, I contacted flight control for our mission assignment. We were given a heading, a target area location, along with our FAC's callsign and radio frequency. As I observed the weather around us, we switched radio frequencies and headed for the target area near the city of An Loc, about 50 miles north of Bien Hoa.

VMA-211 A-4E with twelve *Mark 82* 500-pound bombs enroute to a target. (www. arcforams. com)

Nearing the target area, the skies opened, and in clear and bright conditions we contacted the FAC and got a target brief. We spotted the FAC orbiting south of the target area above our altitude, a sign of possible SAM activity. Today's target was a walled enemy-occupied village with its main gate opening onto a major north-south road. I was struck by how much the village looked like one of the circular practice targets we had used on the Yuma, Arizona and Fallon, Nevada ranges. I turned on the gunsight.

Dive bombing with an A-4 was part science and part art, with an added measure of luck. It required practice. The gunsight was only one generation removed from its Second World War predecessors; there was no radar or computer to improve bombing accuracy. It had nothing like today's attack airplanes with their sophisticated sensors to provide steering to the target and to calculate a weapon release point. For us to bomb accurately, we had to accomplish five things simultaneously: The gunsight aiming point (pipper) had to cross the specific target precisely with the airplane at the calculated dive angle, on the planned airspeed, corrected for wind effects, and at the correct release altitude. As I said, this took practice, lots of practice.

I made sure that the bomb arming and ordnance release switches were set properly. As I dived toward the target, the FAC cleared me to drop the first bomb at any target I wanted inside the circle, and I flicked on the Master Arm switch. I realized that because he didn't specify a particular target, the FAC didn't expect much from us. Since the friendly troops were well clear of the target, he was not concerned with a precise target selection. What I saw through the windscreen seemed like a practice bombing mission back home, and I aimed for one specific building at the dead center of the target circle. I quickly crosschecked the aircraft altitude, dive angle, airspeed, wind correction, and gunsight pipper placement, and at the release point as I thumbed the bomb release button, I felt a 500-pounder eject from its bomb rack. Pulling off target at four "Gs," I looked over my shoulder to see that the bomb impacted about 75 feet from where I had intended it to hit, but I felt pretty good about the accuracy of the drop, knowing my aim would improve with the eleven remaining bombs I was carrying.

This mission began to take on all the qualities of a "milk run." Below us was a big, circular target, there was no enemy opposition, the friendlies were positioned well clear, and the weather became no factor to accomplishing the mission. Maybe it was planned that way for a couple of new guys to achieve success and to develop a sense of confidence.

Woody rolled in next and the FAC cleared him to "drop another one anywhere inside the circle." He didn't get what he expected; Woody's first bomb totally missed the circle and landed outside the village's protective wall and in the ditch across the road from the hamlet's main gate. I immediately thought, "This can't be good." Disappointed with Woody's bombing inaccuracy, the FAC began to chide Marines in general and Woody in particular. I remember him radioing us, "Can't you jar heads do any better than that? Geeze, what a couple of rookies." Then he interrupted his harangue to say that the ground commander was calling him on another radio told us to hold and wait for him to finish.

I was embarrassed; Woody was humiliated; our personal reputations, our squadron's prestige, and the Marine Corps' repute was on the line. Was the ground commander's radio call to the FAC bad news that we injured friendlies by mistake? We didn't need rebuking from the FAC, (especially an Air Force guy), but we held, waited, and hoped.

Shortly he came back on the radio and I noticed his voice had changed, "OK guys. The ground commander reported that Lead's bomb was effective with several enemy casualties. That got the remaining bad guys up and running. They ran out the front gate and jumped into the ditch across the road, just where Dash Two's bomb landed. Great job, Marines! You guys do excellent work!"

Sometimes, as the old saw in Naval Aviation goes, "It's better to be lucky than good!"

Ft Belvoir Diary

By Richard DeLisle

I SPENT MOST OF MY DUTY TIME at Ft. Belvoir, VA, just outside of DC. There was always something to see or do on weekends.

One of my most memorable weeks led up to JFK's inauguration on January 20, 1961. We were awakened early by noise in the street outside of our barracks. Two Airborne units were being housed on our base because they were going to serve as Honor Guards along the parade route. They were running down our street and yelling cadence. It was cold out, but they were in shorts and T shirts.

Two nights before the Inauguration, I decided to go down to the beerhall and have a beer. When I got there, I saw that it was completely filled with Airborne units. More than 150 souls. I decided to have a beer at the bar and leave. There was an Airborne guy standing at the bar too and we struck up a conversation.

Before long, some fights broke out and I decided to leave. The Airborne guy told me to stay put and I would be safe. The fights were between the two Airborne units. Soon, our base MP's were called to end the fights and damage. When they came in, they were promptly escorted out of the beerhall by some of the Airborne. The Airborne guy told me I was about to see something that I would remember for the rest of my life.

Four Airborne MP's came into the beerhall. They were big and their uniforms were straight out of Hollywood. Think about the movie, "A Few Good Men." Four against one hundred and fifty! When they walked in, a hush set in over the beerhall. The MP in charge said, "I want everyone out of this beerhall now!"

One guy said, "I just got my beer. I'll leave when I finish it." The lead MP walked over to his table and slammed his baton down on the table spilling the beer. He said, "I want you out first and if you need help finding the door, I will help you."

The Airborne guy said, "We had better leave now. These guys have no rules and will bring order any way they see fit. Guys can really get hurt with those batons and then they would also face punishment for their actions." The place was emptied in no time.

The night before the inauguration, there was a major winter storm in DC. It threatened the Inauguration because the streets were filled with eight inches of snow which is gridlock for DC. The Army Corps of Engineers at our base was mobilized and was charged with clearing all the streets along the parade route. They did a great job and the Inauguration was able to go on.

The Airborne units stood along the parade route as did many other soldiers. There were an estimated 16,000 military present at the Parade. The Inauguration went off without a hitch and, like the Airborne guy said, I never forgot the night at the beerhall.

*****

When you spend so much time with other guys, in close contact, you develop memories that last forever. Some that I remember are:

<u>Marv</u>

Marv was a self-professed hillbilly from West Virginia. My mother was born in West Virginia, so Marv and I developed a close bond. Some of my strong memories of Marv are:

Marv knew that I was a basketball player and he loved the sport too. He told me that I had to go home with him one weekend to see a player at West Virginia University. The player was Jerry West and if you know basketball, you know Jerry West. I will never forget that weekend that I saw Jerry play. He was incredible and went on to be a phenomenal professional player.

Marv's family lived on the side of a mountain in the West Virginia hills and his dad worked in the coal mines. Whenever Marv went home for a weekend he always brought back a jar of "white lightning." Very powerful! Marv also introduced us to chewing tobacco.

We had rabbit for supper that his dad shot. It was very good.

After we got out of the service, Marv got a place at the Pennsylvania state line and he was living off of the land. Selling fruit, vegetables, firewood and trees for lumber. We went to visit him and we really enjoyed the stay and how his family was making a living.

Contractor

I am sorry to say that I do not remember this guy's name, only his actions He was a contractor in his out of service life. He was in the National Guard and had been called up to serve six months of active duty. He worked out a deal with our brass and he took on contract jobs all over the area. We never saw him except at the end of the day. He offered for pay work to all of us, so many people made extra money. We were sure that he was compensating the staff so he could ask for someone to be excused for duty for a day so they could work for him. He made quite a bit of money during his six months and other soldiers did too.

Marshall

Marshall was from Tennessee and he showed us something one evening that most of us had never seen before. After the lights went out it was pretty dark in the barracks. We were all in double bunks and Marshall was in a top bunk.

He said, "Listen up you Yankee boys and I am going to show you something you have never seen before." He lit his lighter and held it to his butt. He blew a fart and a stream of fire shot out of his rear! He said, "I'll bet that none of you knew that farts burn."

We couldn't stop laughing for a long time.

By the way, Marshall was selected to go off to Officer Candidate School.

The Upside-Down Spin

By Charles F. Willard

LIEUTENANT DON HEINDEL was my first instructor in Air Force pilot training. Don was sharp, laid back but not cocky. It was more an aura of calm competence despite the unusual and sometimes dangerous circumstances we fledgling student flyers exposed him to.

He had a more-or-less gentle, sometimes mildly sarcastic approach to student correction.

Of my four instructors in pilot training, only Heindel showed me some advanced procedures not sanctioned by the syllabus. He alleged that these were for my safety.

"Willard, given the off-the-wall—verging on bizarre—flying methods you employ, you may 'just possibly' require knowledge beyond the usual recovery techniques if you are to have any chance of surviving this pilot training experience."

I told you he was "mildly sarcastic." I thought he was kidding. He proceeded to show me he wasn't.

"I'm going to show you an inverted spin. Hopefully, we can recover from it."

Now, I started to be concerned. The airplane can begin to "spin" after it "stalls." A "stall" is when there is not enough air passing over the wings to keep them doing their job of supporting the airplane in flight. The aircraft is said to "stall," and it starts to fall out of the sky, usually by spinning round and round as it descends. There is a technique to recover from this situation, but it requires skill and proper use of the controls. My recovery technique from a regular spin wasn't the best. I couldn't even imagine an upside-down one. We were all instructed that if we got ourselves into a spin and had not recovered from it by the time the aircraft descended through 10,000 feet, we were to eject: i. e., "punch out," "hit the silk," "egress the aircraft."

While a right-side-up spin in the T-37 training aircraft is hairy enough—the world goes up and down and around and around like a whirly jig while the altimeter unwinds from 24,000 feet, where we start, toward the punch-out number of 10,000 feet faster than your dizzy and addled senses can absorb—an inverted spin will increase that feeling of helpless, sickening confusion by many times.

"We enter the inverted spin from a right-side-up one," Heindel said. "We do this because, if you over-correct in your attempt to recover from the regular spin, you can find yourself in an upside-down one."

I was listening with rapt attention now. This did not sound like it was going to be a pleasant experience.

He continued, "You will find yourself flipping upside down, still spinning. You will pull about two negative Gs as you flip. This will make you nauseous, send all your blood into your skull, and the years of residue that has collected on the cockpit floor from hundreds of flights—filth from flying boots, student upchucks, ashes and butts from illicit smoking in the cockpit—will spray into your face. Your breakfast will start back up your swallow pipe. You will be spinning inverted, with the sky below and the ground above you. You will wobble up and down and round and round, hanging from your shoulder straps, grit in your eyes, the instruments doing weird things. You must keep a clear head, or you will crash and burn."

Mein Gott im Himmel! I thought to myself. It was an expression I heard our contingent of German student pilots use under dire circumstances. *Heindel's gonna make me go through this?*

He did. He put us in a regular spin and got us inverted by firewalling the stick and holding it there too long. This flipped us upside-down, and all the bad stuff he mentioned happened. My teeth were floating in bile. Dirt and filth were everywhere. I could feel myself starting to "red out." This is where your blood goes into your head, and you literally "see red." We hung from our shoulder straps, watching the ground oscillate above us. As the uncanny gyrations reached a regular cadence, Heindel pulled the stick back into his belly and punched a rudder pedal, and the plane "split S-ed" into level, right-side-up flight. The bile receded, the filth fell back where it belonged, and my vision cleared. We headed for home.

"Whew! I guess that's a good reason to do a better job on my regular spin recovery procedure." I said, still shaky from Heindel's little demo.

"Ya think?"

"Yes, I do."

Blessèd Toys

By Vaughn Stelzenmuller

I SUPPOSE PEOPLE THINK an Army infantryman would have mostly tangible thoughts, if he thought about things much at all. *Intangible* things, like differential equations, or pre-Socratic world ethics, or monetary control of floating currencies probably lie outside his daily interests. But tangible objects, yes indeed. Say a string of various rifles, a good outdoor session at a PT course, or that great club downtown. Now that's tangible.

My favorite was clearly tangible. *Oh*, not a rifle. As much as I did like the M16 of the late 60s, I hated cleaning the damn thing. It was lighter than the M1 I had trained with earlier. You could carry oodles of ammunition because of the relatively slender 7.62mm rounds. If you needed to put an enemy out of action, that round would pack considerable punch. It fit the shoulder better than earlier Army rifles. But cleaning the thing was hateful. That spring! That damn recoil spring. In those times you had to watch that piece of jacal crap closely on disassembly, or you could be defined as an unarmed target very quickly by misplacing or mis-assembling the part. Cussing would not get it back together, even in a pinch.

The M1 rifle, heavier though it be, did not have this problem. You couldn't make M16 mistakes with it, and you couldn't get hurt disassembling or reassembling; only if getting the thumb or finger caught in the breach when doing "inspection arms" sloppily. Cussing *will* get the thing back in order, with or without combat or parade stress.

Nope. I would not pick a rifle as my toy. If I could choose something really neat, it would be a truck.

There is nothing on the road neater than the Army M37 Dodge 3/4-ton 4X4 truck. Bigger than a jeep, smaller than the drab deuce-and-a-half two ton truck, that baby was just the right size. Hell, I probably couldn't even drive one! Mike Hill, our veteran friend who passed along to the Great Motor Pool in the Sky a couple years ago, is smiling down at me right now with my choice. That M37 could fit just the right amount of cargo, or a squad of infantry to get somewhere really fast. It was trim and sexy. Its big fault came from its manufacturer's name itself, Dodge: it really was *dodgy*. Maintenance on the thing at the time was a horrible cost. But it was truly a cool vehicle. Mike would agree.

A lieutenant at Fort Hood (Texas) gets the blame for cementing the love of this truck in my own memory, though. The story goes (I never met the man myself) that he hopped in a 3QT truck from some motor pool, packed a carton of C-rations in the back, and hauled off down range somewhere on the humongous post for an extended stay. They said he was American Indian, and likely had got fed up with Army life. Of course, he did get tracked down, though it was likely difficult because of the size of Ft Hood.

Maybe the idea of the man taking off like that, world and future court marshal be damned, made a kind of man-truck story into a permanent romance tale. Imagine this AWOL soldier: a good-ole' American man following his Native American roots, whatever his tribal background, and just heading out into the immense wild by himself to commune with that great *wild* itself. Even a carton box of C-rations would taste like banquets with all this freedom. Here he probably sat, watching a sunset and sunrise in the U. S. southwest, with maybe some prairie animals scooting to-'n-fro, munching on C-ration "white bread" from a can. Nothing else here to do but look at the sweeping beauty that Nature refused to let him ignore, and think about everything—or blessedly nothing.

The lucky S.O.B.

Collision Course

By Charles F. Willard

DR. WILLIAM PEARSON TOLLEY served as the seventh Chancellor of Syracuse University from 1942 until his retirement in 1969. He stood out as a leader of people. They said he "lived in the future," a step ahead in seeing and preparing for what was coming. Charming, confident, optimistic, and clear-headed, he had a knack for developing consensus around his ideas for what must be done next.

He became nationally known as an innovative, visionary champion of higher education for anyone who aspired to it and for adapting the University environment to meet the shifting needs of the times. This was especially true for veterans. Chancellor Tolley was recognized by the government for his academic leadership, which resulted in his having a strong hand in drafting "The Servicemen's Readjustment Act of 1944," known to us all as "The GI Bill." The greater Syracuse community revered him for his contributions and accomplishments.

It was my good fortune to attend S. U. during his tenure. I joined Air Force ROTC as a freshman, and after four years, I found myself chosen as the cadet commander of the AFROTC Corps. One of the responsibilities of this "exulted" position is to lead the Corps in "drill and ceremonies," i. e., marching. What? The Air Force marches? You bet they do. I still have my Air Force Manual 50-14, "Drill and Ceremonies."

"Why do we need "drill and ceremonies" in the Air Force? The answer involves tradition, public opinion and morale, discipline, leadership training... etc., etc., etc." Unquote.

During my four years at S. U., I spent every Tuesday afternoon between one and two PM in marching practice with the Corps. Each cadet had to take turns directing the drill. The first couple of years, I wasn't very good at it. (Understatement.) Even with the instructor calling out, "hut, twoop, thareep, horp!" I couldn't catch the rhythm.

Two left feet wasn't the half of it. I would give commands on the wrong foot, march the whole group off of curbs and into the bushes, then forget the correct command so that half the corps went one way and half another, usually banging into each other and sometimes knocking someone down in the collision.

I don't remember the secret, but things finally began to gel. I got rhythm! From there, I became pretty darn good at it and even started to enjoy it. By the time my senior year rolled around, I thought we might inject some entertainment into the weekly drills by trying some fancy stuff—commands and maneuvers that were in the book but had never been part of the basic instruction routines. We made it fun, and the younger guys, the underclassmen, caught on too.

Graduation week was approaching. This was a big deal at Syracuse during Chancellor Tolley's time. It brought a crescendo of rituals and celebrations, concluding with the graduation ceremony in Archbold Stadium the following Sunday. It was a week to celebrate the successful culmination of four years of student and faculty hard work and achievement.

The kick-off event was the Chancellor's Review, held the Sunday before graduation on the main quadrangle, and it spared no bells and whistles. This Parade and Review involved both the Army and Air Force ROTC Corps and the Syracuse Marching Band, in all, a contingent of several hundred marchers. Not quite the Rose Parade, but impressive, nonetheless. The ROTC units led the procession, with the band following so that all could hear the music, and the band would finish with a performance in front of the reviewing stand, serenading the Chancellor, his staff, and those in the stands about to receive awards.

The Parade and Review lead unit alternates yearly between the Army ROTC and the Air Force ROTC. As fate would have it, this was the Air Force's year, which meant that it fell to me to be the guy up front, giving the commands about what to do when, as I led the whole shebang.

Talk about sleepless nights! There was no time or provision to gather the combined marchers in a rehearsal. I prayed I could get through this without snafus or regression to the arhythmical "two left feet" days.

An hour before step-off time, we began to get the whole contingent formed. It stretched hundreds of feet from the main quadrangle far back into the shadows between the buildings and the back alleys. The band played a couple of warm-up numbers from somewhere between the hallowed halls, but the echoes made it hard to catch the primary notes. It was a warm day. Sweat was pouring from under my "fifty-mission-crush" hat. It came time to step off. Here goes nothin'.

The parade route was laid out with flagged wooden stakes at its edges. These marked where I had to execute the "Column Right" and "Column Left" commands to negotiate the marching route so that we passed in front of the reviewing stand. We approached the flag where I had to give the final "Column Left." I was marching backward, looking over the marchers, ensuring everything was lined up correctly. It all looked good. The band had come from behind the buildings, and everyone was in step with the rousing Sousa march. At the last stake, I gave the "Column Left" command (on the correct foot, too). Man, we looked good.

I executed a snappy "in step" about face, now marching forward, not losing a beat—and found I was marching the entire Corps straight at Chancellor Tolley and his staff. They were positioned a few yards in front of the review stand, absolutely where they should be. Whoever had laid out that last stake must not have realized the Chancellor and his little group would be out front. They had placed the stake so the parade would pass nicely in front of the stand, with no allowance for anyone in front of it.

Standing on crutches due to a recent accident, the Chancellor turned his head toward us and realized what was happening. With nowhere to go and no time to get there, he and his group helplessly held their positions. They were "dead ducks" unless I did something fast. The whole Corps was bearing down on them, and I was leading the steamroller. I had less than twenty-five yards to fix this—at 120 paces a minute, not ten seconds.

Thank God for all that drilling. As though it were all planned this way, I shouted, "Column, Half Left, March!" I pointed left so there would be no confusion in the ranks. Then, clear of the trouble, "Column, Half Right, March." Everyone followed. We looked good. Then, as we approached a relieved Chancellor and his group, "Eyes, Right!" Looking sharp, we pulled it off, but not without a hitch.

Down Time Ditherings

By Richard Terboss

EVERYONE'S LIFE IS PUNCTUATED with what you might call "down time." Sometimes that down time lends itself to action where your feet do not move but your thought process does. It may be only for a few minutes, but it happens. What follows are results of the above. These two vignettes could easily apply to in-service time as well as post-service time. Amazingly, I've discovered there are considerable process, content and intent overlaps. Next up is an example of Basic Training daydreaming, and beyond. Let's call it *The Big Picture.* It is followed by a mood descriptor that will sound familiar to us all. This one is entitled *A Disgrunted Person's Lament.*

THE BIG PICTURE

You need to rest, my concerned mind says,

Put down your textbooks, your encyclopedias, and your scripture.

I cannot do that, I reply.

That negates answers I seek to THE BIG PICTURE.

As soon as something enters the senses,

I ask does that fit inside the BIG PICTURE'S fences?

Competing visuals and thoughts make it difficult to respond.

Pick one... it's the best you can do.

Then put a thought thumbtack through.

And move on.

I must renew my search for answers to the BIG PICTURE'S questions.

Each of which can turn your brain into fuzz.

But I've narrowed the search down, before my brain becomes unwound.

To consider, simply, **All that is… All that is not… and All that once was.**

A DISGRUNTED PERSON'S LAMENT

I have to get out of here,
And somehow go away and get wild,
Shed all my concerns, but not my clothes… probably,
Otherwise be like an adventurous child.
Tomorrow, as I do my job; in-service or out,
A DISGRUNTLING thought may interrupt my task.
But who has not smiled when reminded
To be patient because "this too, shall pass."

DANGER! CELEBRITY ON BOARD!

Live From Kabul!

By Tim Hansen

IT WAS ANOTHER DAY at the New Kabul Compound as officers, NCOs and enlisted scurried off to staff meetings, answered phone calls, or stared intensely at screens, giving the appearance of looking busy. In my own daily deluge of emails, I noticed a note from the ISAF PAO - more specifically, from a young Marine Captain by the name of Tony Banks. It was another request for a media embed but he needed my response right away. It was March 2013, fighting with the Taliban and other discontents now happened in the outlier provinces of Afghanistan, but interest in logistics suddenly shot into the media stratosphere because President Obama ordered the exit of U.S. forces from the country. Bagram Air Base had become one giant wash rack for vehicles heading back to the U.S. Who would be interested in that? I wondered. I reached for my phone.

"First TSC Public Affairs, Lt. Col. Hansen," I said tersely.

"Hey, sir. Got a request here for a news crew from AXS TV out of New York to spend a few days covering the withdrawal from Afghanistan," he replied. "Thought you might be interested."

"Thanks for thinking of us, Tony. Who is the reporter?"

"Ah, let me see ... Dan Rather."

"Dan Rather? You mean Dan Rather who reported on the Vietnam War and stood up to Richard Nixon during a press conference?"

"Yessir, and I need to reply back no later than noon today."

I stopped skimming and sorting through my emails, one of my multitasking skills as a staff officer, and saw something momentous and if I didn't take the shot, I'd never forgive myself for passing up a great opportunity to tell the story of the First Theater Sustainment Command's role in the withdrawal of U.S. forces from Afghanistan.

"We'll take it, Tony. Send me the details," I said reflexively.

"You're sure, sir?" Tony asked, surprised by my near immediate response. "You want to check with your boss and call me right back?"

"Positive," I shot back. "General Steck will agree to this embed. No worries there, but I will need your help with transportation since, as you know, we are spread out a bit."

"You got it, sir. I know we can lay on a Black Hawk at the least."

"Thanks, Tony. I will let the general know right now and will send you an itinerary."

I sat at my desk staring blankly into my screen, trying to process the phone call I just had. How in my lucky press releases did this happen to me?

Instinctively, I shot up from my desk and headed to General Steck's office down the hall. His door was open, his light was on, and his bulldog chief of staff was nowhere to be seen. I couldn't believe how the stars aligned for me on this one. I stood at Steck's door and found him furiously writing out his notes on 3x5 index cards. I liked this guy. He was smart, snarky, and Jewish. He paused from his writing and glanced up at me.

"Hansen. What do you have for me?" he asked.

"Sir, would you be interested in doing an interview with a reporter from New York?" I asked.

"Who is the reporter?" he shot back. Steck was good with the media and reporters liked him. "I don't want another interview with *The Yonkers Tribune*."

"Well, sir, how about ... Dan Rather?" I suggested cautiously.

"Seriously," he replied as he locked his gaze on me.

"Sir, you could chat with him on Friday ... if that fits into your schedule."

"Why are you still standing there," he retorted and grinned. "Check with Colonel D'Angelo and get him on my calendar."

I knew Steck would approve and went right to work, drawing up a list and contacting other support commanders who might want to interview and give a tour of their units. All agreed and I had my itinerary completed by that afternoon. I needed transportation for Dan and his crew of three, his producer, his cameraman, and his sound technician. I called Captain Banks over at ISAF.

"Hi, Tony. Just shot you our itinerary. General Steck is good with interviewing with Rather on Friday," I said. "But, I need transportation to get Dan and his crew from Kabul up to Bagram. Anything available?"

"Let me check, sir," he replied. "Hmm ... of all days, everything is tied up. Wait. I know the pilots who fly General Dunning's G3 and they might be heading up there to pick up some of the ISAF staff. I could ask them for a favor."

"Wait ... are you saying you could lay on the ISAF commander's Gulfstream jet?" I asked.

"I'll call you back with an answer in 15 minutes, sir," he asserted and hung up.

True to his Marine Corps disposition, Banks called back within 15 minutes and confirmed the Gulfstream.

My head was spinning to say the least, but I managed to pull together a plan that any news outlet worth its weight in videocams would salivate over.

Those intervening days became a blur, and before I knew it, I found myself standing outside of our headquarters to welcome Dan Rather and his AXS-TV crew. Was this really happening?

"Hello, Mr. Rather, and welcome to the New Kabul Compound," I said and extended my hand.

"You must be the Lieutenant Colonel Hansen we have heard so much about," he said disarmingly. "And please, call me Dan."

I should have known he probably knew all about me with the millions of sources at his disposal.

"I think you will enjoy meeting the soldiers in this unit, Dan, and we have a lot to cover," I replied. "We will be heading up to Bagram this evening, but first, I'd like to introduce to my boss, Gen. Steck."

"Lead the way, colonel," he said enthusiastically.

Yikes! I thought to myself.

When we entered the building, soldiers stopped and acknowledged him and his crew. I was amazed that millennials recognized him. We finally made it to the third-floor conference room and, as if right on cue, Gen. Steck came in to speak with him. Even I was impressed.

Next, we headed to the military side of Kabul International Airport. Banks had arranged everything perfectly and we made it to the ramp without a hitch.

We walked out onto the apron toward the sleek Gulfstream jet. It had a stunning beauty and grace -- all its own. The auxiliary engine was already humming and a charming female E6 waited at the top of the ramp to greet us.

"Hello, Mr. Rather, welcome to Afghanistan," she said as he entered the jet.

We sat down in soft leather seats and I stared out of the wide oval window as the amber and orange hues of the Afghan sunset faded behind the majestic peaks of the Hindu Kush. I sat opposite Dan and we started talking. The female E6 came up to us for what I thought would be to remind us to fasten our seat belts.

"Would you like some chocolate chip cookies, Mr. Rather?" she asked. "I just made them and they're fresh out of the oven."

"You baked chocolate chip cookies on a Gulfstream?" I asked, aghast in disbelief.

"Sir, this is the Air Force," she replied.

I simply couldn't believe it, but her cookies were delicious and I engaged in a conversation with Dan about everything from ancient history, to reading Homer to his grandchildren, to golfing in Texas. I realized then and there I was in the company of a master storyteller and the reason he is who he is.

In no time, we were on the ground at Bagram and our odyssey was about to begin.

Farmer

By Charles F. Willard

WE HAD BEEN FLYING up and down Europe from Narvik, in the far north of Norway, two degrees above the Arctic Circle, to Naples in Italy, a much warmer clime, with many ups and downs in between.

The mission was Operation Big Lift, a huge multi-service, multi-nation logistical exercise designed to show the Soviets that whatever aggressive cross-Iron Curtain move they might be planning, NATO could counter it with quick and massive movements of troops and armor.

We had a basic five-man crew today—two pilots, a navigator, a flight engineer, and a loadmaster—so our crew duty day was twelve hours. We had been flying this mission for two weeks, taking different C-130E Hercules aircraft and mixing different crews, whoever and whatever was available.

Today I was flying with Captain Carroll Moran, who was relatively new to our squadron. I had not flown with him before.

He was an Arkansan, born and brought up, and he had a country background similar to mine. He looked like a country boy—a little portly (all those chitlins, I supposed) with a thick, shaggy mop of reddish-brown hair protruding in all directions from under his fifty-mission-crush cap.

He had several thousand hours of C-130 time, starting with the A-model and progressing through B's to our new E-models. The experience showed. He was smooth and precise in the air and every landing was a "grease job."

This contrasted with my technique, which at that time, was "get it close to the ground, chop the power, and plunk it onto the runway." Sometimes it was "ker-plunk-lunk-lunk."

'Kind of humiliating, flying with a skilled pilot like Carroll Moran.

At the start of the day when we crew were introducing ourselves to each other, the first words out of his mouth were "Call me Farmer." I guess if my first name was Carroll, I'd want to be called something else, too, for obvious reasons—It was sort of like Shel Silverstein's song made famous by Johnny Cash about a boy named Sue. So, "Farmer" it was.

We were at the end of our duty day now, with plenty of ups and downs on our zig-zag flight plan, enough to contrast our disparate landing styles, and our final stop today was Sembach, West Germany. (This was October 1963 and Germany was still divided. Remember that the Berlin Wall did not fall until 1989, and Germany was not united until October 1990.)

We landed (kerplunk), and had the crew bus take us three officers directly to the BOQ. We were sweaty, tired, and hungry after the long day and needed a clean-up. We signed in and they gave us a three-bunk room—made for easier alerting in the morning. The bunks looked comfortable, so no complaints.

The navigator and I wandered off to the latrine for a shower. It was already crowded with other crews. There were about ten shower heads, so our wait was not long. We got cleaned up and ambled back to the room. Farmer was not there.

While we were dressing a bald-as-a-bowling ball, fat fellow, wrapped only in a towel, came plodding through the door, his big flat feet flapping on the tile floor. Our navigator, a crusty old guy, yelled, "Get outta here, you fat SOB. This ain't your room!"

But it was Farmer!

He had left his frowsy, floppy toupee sprawled on the dresser.

Glenn Ford and the Ouija Board

By John P. Bland

BACK IN THE EARLY- TO MID-70S, I was assigned as a flight instructor in the Naval Air Training Command at Kingsville, Texas. My job was to train and evaluate would-be Naval Aviators in advanced flight training, their last phases before their graduation, getting their wings, and receiving a fleet assignment. Their very last challenge was, "the boat," learning to land and takeoff on the training carrier USS *Lexington*, an actual WW II aircraft carrier based in Pensacola, Florida.

During one at sea training evolution, The Mirisch Corporation was on board *Lexington* filming the action movie, *Midway*, an epic war movie which chronicles the Battle of Midway, a turning point in the Pacific Theater of Operations of WW II. Many of the cast members were aboard, including actor Glenn Ford who played Rear Admiral Raymond Spruance. Ford had combined his film career with military service; he had retired in 1970 as a Captain in the US Navy Reserve specializing in public affairs (PAO). As a PAO he had minimal knowledge of aircraft operations at sea.

Actor **Glenn Ford (left) as RAdm Raymond Spruance in *Midway***

Let's digress a bit. In Flight Deck Control of any US Navy aircraft carrier is the ship's Ouija board. This isn't a wooden board for asking Aunt Bessie how she's doing in the hereafter, but a two-dimensional mockup of the carrier's flight deck or hanger bay where the Aircraft Handler figures out how to fit and move about aircraft and other bulky equipment in the confined spaces.

The Ouija board system has worked very well since it was invented during WW II. The board, a replica of the carrier's flight deck and hangar deck, was on a scale of 1/16 inch to one foot. It is about six feet long and two and a half feet wide, about the size of a large coffee

The Ouija Board

table, with the flight deck on top and the hangar bay underneath, like a second shelf. Scattered over both surfaces are small templates representing aircraft, made to the same scale, and located on the Ouija Board exactly where they are located on the flight deck or hangar deck. Colored pins, metal nuts, and washers are placed on the numbered aircraft models to represent items of information (fuel status, maintenance needs, etc.), which relate to that aircraft.

The Ouija board is the centerpiece of Flight Deck Control, located on the flight deck level of the "island," the structure that towers above the carrier deck, starboard and amidships. It is one of the busiest spaces on the ship during flight operations, with people constantly moving in and out of the compartment. It is also one of the most interesting and fascinating places for visitors to see. I was in the busy compartment and standing in a corner out of the way while waiting to man up an aircraft for a flight home.

Glenn Ford strode into the compartment. He was touring *Lexington* to get a feel for his role as a WW II admiral and "to press the flesh" with the ship's company, his fellow sailors, and perhaps one or two adoring fans. Flight Deck Control was hot and humid with water condensing on the ceiling. The air conditioning was doing its best, but it could not keep up with the demand for cold, dry air. Drips of water were falling onto the Ouija board and the sailors were trying to soak up the drips as best they could. One petty officer pushed a folded paper towel across the Ouija board toward Glenn Ford to catch a drip. Ford picked up the paper towel, pulled out his ballpoint pen, and asked the sailor to whom he was to make out the autograph.

Cat Nipped

By Cindy Blair

AS A PETTY OFFICER IN THE NAVY I was not used to insubordination and my cat would not cooperate. I wanted to put her in the brig.

I crept slowly towards the stairs. My breath came in quiet spurts, not quite gasping so I wouldn't make a sound. The object of my stalking lay curled up contentedly on the pillow; never dreaming of what needed to be done. I made it to the end of the bed when; Wham! My cat Sammy flew off the pillow and onto my arm starting rivers of blood with her claws. I rushed to get Band-Aids and antiseptic. How would I catch her now?

Sammy had settled in nicely during the three months since I adopted her, but today was her first veterinary check-up after her rescue. I believe she sensed that something was up and she wasn't going to like it. Poor Sammy! It made my heart break to see her now trembling in fear underneath the bed. "Okay," I thought, "how will I ever get her into her carrier?" Then I thought, "I know what will do the trick!" I put the open carrier next to the bed. I called softly, "Sammy, do you want some treats?" and I put a few chicken flavored treats into the crate.

Unh-uh! Sammy was too smart for me. She knew the reason for the carrier. That meant we were going to go in the car and she hates car rides. I continued to cajole to her. "Come on Sammy, I won't hurt you," (even though I had thought about knocking her out after she had dug her claws into my arm). No doing, she had slunk way back into the corner under the bed where I couldn't reach her. I was forced to call the vet and reschedule for the afternoon. Luckily they had an opening. After hanging up the phone I placed Sammy's favorite blanket on the floor next to the carrier and I put a little catnip on the corner. Behold Sammy crept out onto the blanket. Success! Sammy started purring and rolling around in the catnip.

Quickly, I scooped her up in the blanket, my heart racing wildly. I moved toward the carrier. Sammy started thrashing around violently and yowling as if she was possessed. I almost dropped her, but I hung on as tight as I could without hurting her. I threw her down into the crate and securely latched it. I drove Sammy, howling all the way, to the veterinarian. Wouldn't you know it, once we got there she was as calm as could be. Until the ride home when she was yowling again. "Sammy," I said, "no more catnip for you"!

Previously published in "Afterwords," an anthology from the Veterans Writing Workshop 2020 Hudson, NY

Toby Keith

By John Steele

WHEN I WAS IN IRAQ back in 2011, country music star Toby Keith came to our little corner of the world. Toby Keith did an awful lot for military members. It had been nine and a half years since 9/11, yet it was Toby's eleventh USO trip to Iraq and/or Afghanistan, and the eighth time he had played in Baghdad.

The concert was scheduled to start at 2000. It was at the Liberty Post Exchange (PX) stage. We were on Camp Victory, but the two FOB's (Forward Operating Bases) were connected. We wanted to get there early. Something we didn't expect to see there was tail-gating.

We got there about two hours early, and there were already several hundred people jockeying for position in front of the stage. Many had their folding camp chairs, and most of them had coolers full of what we regularly call "near beer," non-alcoholic beer. Now I know some people like the taste of beer, but I think more people drink beer for the effects of the alcohol. When the percentage of alcohol is minimal, the effects should be minimal as well. But that didn't stop these people. Almost everyone who was there early was pounding the near beer, acting like they were inebriated, and having a grand ole' time.

There were three couches plopped right in front of the outdoor stage, full of Air Force personnel. It begs the question: Where did they get three couches from? Whose couches were they? How did they get them there? How long had they been there? Those are some of life's little mysteries. Being that they were Air Force personnel, we were pretty sure that they did not carry them there, and they sure didn't bully them away from anyone. The best explanation that we could come up with is a crew chief for a Blackhawk was one of the personnel on one of the couches, and he had one or more helicopters move the couches into place.

At about 1950 (7:50PM for you civilians), a few of us from my unit situated ourselves about 10 feet from the corner of the stage. We were close to the speakers, but had a great view of the front of the stage. There had to have been 2000+ Army, Marines, Navy, Air Force, and Coast Guard personnel there. (Why were there Coast Guard personnel stationed in the middle of a desert? Another of life's mysteries.) Everyone of course had their weapons with them. It was kinda' surreal for some of us northerners to have everyone walking around a concert with either a rifle or a handgun, but maybe it wasn't all that uncommon down south.

At about 2005 (8:05PM), some Sergeant Major took to the stage to give the safety brief. He just started going over what to do in an emergency, and guess what happened? The C-RAM (Counter-Rocket, Artillery and Mortar) started going off. To quote the Sergeant Major, "Really? Are you f***ing kidding me?" Immediately the booms started booming, and the remainder of the 2000 people who hadn't already hit the dirt proceeded to do so. How many times have you gone to a concert, and 2000+ people in the audience are lying face down in the dirt with their hands over their heads? (Besides, perhaps, the later stages of a Grateful Dead show.) Some of the things you could only experience in Baghdad. What an image. Everybody lying face down with their hands over their ears, and their weapons still not touching the sand. What I couldn't see, and I would have loved to, was the reaction of the people on the stage. Toby wasn't on stage yet, but his band members were. I am guessing they were prone too. Shaking? Probably. The points of impact (POI) weren't all that close, but you could feel them and definitely heard them. There were three booms within 5 seconds, and then quiet for about 10 more seconds. Then there was some murmuring and some people started getting up, and Sergeant Major reminded everyone to stay down until the big voice gave the all-clear signal. But the noise of the crowd strengthened until about a minute and a half later when everyone was back up on their feet, cheering for Toby to come out. The Sergeant Major quickly introduced Toby Keith, and out he came. After a small vulgarity laced rant about how they must have known he was coming, and a few more vulgarities about insurgents trying to ruin our good time, and how he wasn't going to let that happen, he went right into his first song. Less than five minutes after an indirect fire attack, Toby Keith started his concert, and all the service members were going wild.

I remember being curious how other entertainers would have handled it. What if Justin Bieber were playing? Would he have come out still? How about Taylor Swift? But the Soldiers seemed to love it, and Toby was into it. About the only thing I could compare it to was back in 1989 when there was an earthquake before game three of the World Series between San Francisco and Oakland, and immediately after the earthquake everyone in the stands started cheering. For us, I think part of it was nervous energy. The POI wasn't on Liberty, but it was somewhere. We hoped everyone was okay. They were. There were no casualties. But we didn't know that at the time. I think the cheering response was our way of saying, "You are not going to beat us, and you are not going to deter us." It was also our way of saying F___ You!

As for the concert, it was great. The sound quality and musicianship of his band was top notch. In his videos, he often had this goofy smile on his face. His shit-eating grin wasn't on display too much that night. I am sure getting shelled had something to do with it, but my other hypothesis is that the amount of goofy smile he had on his face directly correlated with the amount of alcohol he had in his system. It is nearly legendary how he drank in his music videos. Iraq was dry. Toby did talk before one of his songs, "Get Drunk and Be Somebody," about how he had tried repeatedly to get us beer for his concerts. He didn't understand how each year military members could get two beers each for the Superbowl while in Iraq or Afghanistan, but we couldn't get any for one of his concerts. He said that every time he ever talked to a General, a Joint

Chiefs of Staff, the Secretary of Defense, and even in his visits to the White House, he mentioned bringing beer for the troops. He offered to buy it and pay to get it shipped to Iraq, and it was still always turned down. So he told the crowd that we would have to pretend for that song.

When Toby introduced the song, "I Love This Bar," he talked about the restaurant chain that he had called I Love This Bar & Grill. There were about a dozen locations at the time. He rattled off where they were, but the two closest to Rochester were either Foxborough, Massachusetts or Pittsburgh. He pointed west, and said that one was only about 10,000 miles in that direction. He said that if any of us showed up at any of his restaurants, our meals were on him. He did that for years for any active or retired military personnel. So guess what I added to my bucket list. I never did make it to one of his Bar & Grill's though. Most of them closed in 2014 and 2015, and the ones that remained were no longer owned by him and his group. Maybe he gave away a few too many free meals for military members.

Toby did three songs that were definitely crowd favorites, though you probably never heard any of them played on the radio." Weed With Willie" was a very comical song about riding on Willie Nelson's tour bus, and the side effects. There was also a song titled "The Taliban Song," about a normal couple living in the Middle East enjoying a happy life until the Taliban moved in. Part of the chorus has the line, "We'll bid a fair adieu and flip the finger to the Taliban," that was repeated a few times in the song. And every single person in the crowd was singing it loudly while flipping the bird. It was a great team-building exercise. The final crowd favorite song that I loved was titled "The Ballad of Balad." It is about a recruiter who gets a guy to join the Army, and the guy then gets shipped to Balad. Balad was a place just north of Baghdad, we had a pretty big base there. The song talked about some of this new soldier's experiences in Balad. It is another one of Toby's witty and funny little ditties. The last line of the chorus of "The Ballad of Balad" that gets repeated several times in the song is this new soldier telling his recruiter, "Ah, but you never told me I'd get my ass shot at."

Toby also did a lot of his hits. He closed the concert with probably his two best known songs for the military, "American Soldier" and then "Courtesy of the Red, White & Blue."

Everyone I talked to loved the concert. It was a great atmosphere and a chance to hang out with battle buddies and relax. It was a small escape from reality for a little bit. As Toby put it, it was a chance to bring a little bit of home to a shitty place. We were fifteen feet from a musical icon, singing and dancing the night away. If only we could have gotten some real beer. Thank you Toby Keith, and thank you USO.

Know how I wondered how other celebrities would have handled the situation? Well, I didn't have to wait that long to find out. A few weeks later, Bobby Kimball & Friends played at the Camp Victory stage. Bobby Kimball of the band Toto, Bill Champlin of the band Chicago, Dave Jenkins of the band Pablo Cruise, and Jaime Kyle, Dave Jenkins' wife and an accomplished singer/song writer herself, were the four featured musicians. They, along with a few supporting musicians, played some of their hits from the heyday of these three fabulous bands. It was a pretty good show. Mostly. Until the C-RAM went off. And the big voice started announcing, "Incoming! Incoming! Incoming!" And then there was a pretty big boom. The ground shook. The dust settled. And we started getting up again and chanting for the band to come back out. The "All Clear" was given. We continued cheering and chanting for the band. And we waited. We cheered some more. And some more. After several minutes, Jaime Kyle, the female, scampered out in a crouched position to a microphone at the front of the stage to say they weren't ready to come back out." YOU ARE USED TO THIS, WE AREN'T!" She then hurriedly scurried back and ducked out of sight.

Now stop for a minute Mr. Bobby Kimball, Mr. Bill Champlin, and Mr. Dave Jenkins. The three of you are such chicken-shits that you sent Ms. Jaime Kyle out there to tell us you guys were chicken-shit? That is even more chicken-shittier. Dave Jenkins, you had your wife go out there? We could have given you guys a pass if you just didn't want to continue the concert. But then you had the female go out and tell us? What, did you guys draw straws, and she just happened to get the short end of the stick? That is embarrassing on your part. Thanks for coming though. You sounded great. We appreciated you taking time out of your schedule to show your support and bring a little bit of home to us. Sorry it was an abbreviated show. But next time, don't send the female out to tell us, even if she wanted to be the one to go out there. If she is all gung-ho and says she wants to do it, tell her too bad. Don't let her prove she has a bigger pair than you do.

"HANDS ACROSS THE SEA" (OR COUNTRY)

Marlboro Country

By Charles F. Willard

THIS IS A RUSSIAN MILITARY HUMOR STORY—although I doubt they knew how humorous they were.

In December 1985, I was on a Japan Air Lines Boeing 747 from Orly Field in Paris across Siberia, destination Narita Airport in Tokyo.

At about 8 PM local time, the pilot made a scheduled refueling stop at Sheremetevo Aerodrome outside of Moscow. It was a bizarre experience. At that hour on a Friday night, almost any other large city airport in the world would have been teeming with air traffic. Here, we were the only plane in sight. Except for us, the field was deathly quiet.

Since this was a refueling stop, all of us passengers were ushered off the airplane. As we came out of the jetway, we found ourselves faced with a line of Russian troops. Each had his Kalashnikov at the ready. They motioned us forward. We had to stay in single file.

Remember, the Berlin Wall did not fall until 1989, and the USSR did not implode until 1991. This was still the Cold War.

In full combat gear, the soldiers with grim scowls looked menacing as only Russians could. Trepidations ran through our crowd as the soldiers motioned us forward. Can we get back on our plane and get out of here? Or would we be "lost forever 'neath the streets of Moscow, the flight that never returned?"

As we made our way further along the guarded walkway, it became clear what the objective was.

Ahead of us in the distance was a large gift shop. They wanted us to spend our foreign currency on Russian trinkets so they could give us near-worthless rubles in change. A surreal sight greeted us as we rounded the corner and approached the entrance. There, big as life, was a massive poster of a Western scene featuring a cowboy leaning against a coral fence, with the caption in English, "Welcome to Marlboro Country."

I bought a set of those nested dolls and a beaver hat, and within the hour, we were back on the plane and on our Redeye way over Siberia to Japan.

Are You There, Moriarty?

By Vaughn Stelzenmuller

THERE ARE TIMES when one really tires of the uptight, measured, controlled, self-centered, self-satisfied officer corps all around me, *uh,* around *one*. There are times when one can truly enjoy the experience of a through-the-ranks field grade officer, especially when we are all gathered around an afternoon German *gasthaus* for a monthly, *uh,* staff meeting set up by, and run by that same through-the-ranks officer.

The afternoon wears on, gracefully sometimes and less so other times. The gasthaus owner was secretly and less-gracefully cheering us on (silently), bringing multiple mugs to wash down the excellent, excellent Wiener schnitzel lunches he had just served us. Mr. through-the-ranks officer, our *Leader,* is the one we let talk the most, mainly because he (as well as the E7's at the table) had far more service "experience" to talk about, and also because he would take the heat from the corps chief of staff later for our group afternoon of *staff discussion,* far from that intrusive full-colonel's eye.

Now, our G-1 group leader had many life skills, and learned how to be a senior officer with a real-life touch. Including boring Allied military socials. Many senior officers in Europe truly disliked attending those functions, for example, mainly because so many were token events to show off who was the most uptight. Lieutenant colonel G_______, though, taught us that we need to go to these events, and just play along until something "interesting" developed.

What an open-ended and optimistic view of life! "Go, then see what develops!" His story of one Allied dinner, after all the senior-senior officers had tactfully left the party, was one I still tell. Apparently there were mostly Brits left at the event after ten o'clock (and after maybe ten beverages, each), and they picked up a game of *Are you there, Moriarty?*

Rules for *Are you there, Moriarty?*: Two participants join left hands, with one of the participants blindfolded. The blindfolded one gets a rolled up newspaper in the right hand. Mr. Blindfold calls out, "Are you there, Moriarty?!" The other then has to answer "Here!" and then ducks. Mr. Blindfold takes a swing at where the other's head ought to be. If he misses, they change places. For one round.

That's it! Whoever said those Brits were stiff-upper-lippers like the 1800s' Arthur, Lord Wellington never saw these mid-level officers going at it like Welsh footballers.

This story is uproariously funny. No civilian I have told this to has ever either laughed or even gotten the point. Partly because there *is* no point. Just good, clean, abusive, nonlethal, abrasive soldier-boy fun, and likely a typical outlet for the grimmer parts of infantry life. Sort of *Three Stooges* with a real newspaper and a real head.

Ah yes. What always happened between our *Leader,* Colonel G_____ and that chief of staff? After each and every one of these infrequent great afternoons the chief of staff would call our *Leader* upstairs after we got back to HQ, and berate the bejesus out of him for being absent from his office downstairs all afternoon. The berating was all true and deserved, and our *Leader* didn't give three floating damns. What a fabulous leadership example! Besides, he had numerous beers to get rid of via a 40-year-old prostate, and taking the verbal beating was the shortest way out of the lecture.

There is great virtue during those days when we had no cell phones. No one could imagine that chief of staff *ever* joining in with *Are You There, Moriarty.*

Laughable Instances

By Sue Spitulnik

THESE EXPERIENCES STILL MAKE ME LAUGH from when I was a young Air Force wife.

The following four incidents happened in Biloxi, Mississippi. My husband was fresh out of basic wearing one stripe, attending radio school at Keesler AFB at the time, 1972. That would have made him 19 and me 18 years old. Naive kids, as I see things now.

The first incident I remember happened while we were in the game room of our apartment building, where the residents were either over sixty or military couples with no children. A foursome of the over-sixty crowd was playing cards, and one of the men told the others that he had folded his wife's p-a-n-t-i-e-s along with the rest of the laundry. He spelled that word like it was too revealing for our ears or that we had never seen the opposite sex's underwear. Then, like nobody's business, he started talking about rubbers. My husband and I gave each other a look and quickly departed so we didn't laugh out loud. But the joke was on us. They were playing bridge and talking about the rubber hand, which is regular vernacular for any card game. And we thought we knew it all.

In the same building, when two other young Air Force couples and we got together, it was in our apartment. The guys went to school together, and sometimes an extra guy, J. J. , would come over for supper. I don't remember what shenanigans we were up to, but the guys held J. J. down, and we girls covered his neck, front and back with hickeys. Who knew the following day he would get an article fifteen for defacing government property. I recently reprimanded a seventy-something veteran for calling himself stupid when he was young. But that stunt of ours can't be described as anything but stupid. I wonder where those two couples and J. J. are today.

Fireworks were legal in Mississippi, and we often bought Bottle Rockets and had "wars" on the beach after dark. If you don't know, a bottle rocket is a thin ten-inch stick, more of a toothpick, with a firecracker attached to the end. We would put the rocket in a Coke bottle, light the fuse, and then hold the bottle so the rocket flew down the beach at one of our friends. Of course, we never considered someone could get hurt. Generally, one of us returned to the beach in the morning by walking across our apartment building's two football field lengths front yard to cross Highway 10 and picked up the rubble we left. The building is gone now, 2023, and a casino sits in that spot.

One more Biloxi happening. My Airman husband was invited to a Staff Sergeant's home for an afternoon gathering with instructions to "bring the wife." We thought it was a get-acquainted cook-out and were highly impressed someone of that rank had thought to include us. It turned out to be an introduction to Amway. We weren't impressed when we left, and we were still hungry.

Our next base was Plattsburgh, New York, 1972 – 1973. For the first three months, we lived in a four-plex apartment building. We liked to hang out at a bar close to home so we didn't have to drive. I didn't drink at the time, but I enjoyed going out. One of the guys my husband worked with had his girlfriend visit for the weekend, and since money was tight for everyone, Mike and Bonnie stayed with us. We had all left the building together, but Mike and I returned to our place because he had to get up for work in the morning. Mind you, I was seven months pregnant, and my winter coat did not button around my belly. Someone in one of the other apartments saw us, assumed the wrong thing, and told everyone else we must be swingers. REALLY, with me and my huge belly! A few weeks later, I ran into one of the other wives in the parking lot. She was nice enough to tell me why no one spoke to us. The look on my face must have told her how wrong the observer had been. We had a good laugh. She also told us about a trailer court out of town that had an opening, so we moved before our son was born. After Mike and Bonnie got married, they rented in the same trailer court. We rarely saw them.

Another Plattsburgh incident occurred when my husband got to work one day, and his line badge, a unique ID that gave him access to the flight line, was not attached to his coat. He called me at home, and I found it lying in the driveway. All is good. It wasn't lost. That afternoon, the base fuels dispatcher, an African American, got a request for a refuel to be done at top speed. As all the other drivers were already out, he asked S. W. what they should do. They reasoned the security person in the line shack who checked the badges rarely came out of his shack, so S. W. could wear the black guy's badge but purposely put it on backward. This day was one of the rare ones. A guy right out of basic doing his best job checked the badge, and within seconds, S. W. was surrounded by machine guns and in handcuffs. I don't recall how long the plane was delayed getting its refuel and what red tape my husband had to deal with, but he came home in a bad mood and showed me the red rings around his wrists. The following morning he had to see his Officer in Charge, and after he was commended for trying to get the job done, he was admonished to never again forget what color he was. The guys now had a long-lasting story to bring up whenever they felt like getting S. W. 's goat, which was often.

And then we moved to England, 1974 - 1977. I have recounted much of that story in the Rochester Veterans Writing Group's first two books, but there are three more specific incidents. After our daughter was born, we hired a babysitter and went on base to the NCO (Non-commissioned Officers or enlisted members) Club. They had live music on the weekends, and it was much less expensive to drink there. I was the designated driver because I didn't drink. As the driver, I got the same treatment every time we left the base after midnight. The Security Police working the gate wouldn't wave me through until he asked, "Have you been drinking, Ma'am?"

Every time, my answer was the same, "No."

"Would you mind taking a breathalyzer test?"

"Of course not." And each time it would prove I was telling the truth. We finally decided that the guard was told to check the driver of any car with front-end damage, which ours had. And, because of how the S. P. s were scheduled, it was rarely the same person at the gate two weeks in a row, so they didn't recognize the car. It got so we would verbally bet on whether I would get stopped.

Our last New Year's Eve party in England was at a pub in Thetford, Norfolk. I had lost all of my "baby weight" and was wearing a long maroon velvet dress. I don't recall the pub owner's name, but the bartender was Davey, who had been in the Navy. (A reference to Billy Joel's song, Piano Man.) Davey came to our table and asked me to walk with him to the owner, who was standing by a booth near the front windows. They wanted to compare the color of my dress to the old, dusty drapes, discussing if they should do an update. I shouldn't tell you they held my hands and helped me stand in the booth seat to get a good comparison. In your visual, put me back to 21 years old and not afraid to have a good time. The latter part is still true.

Another memorable story from England has to do with the difference in televised humor that was allowed. Every UK male comedian at the time, in the mid-1970s, did a skit in female clothing. Sometimes, they wore a wig, so the cross-dressing was complete. My husband was at work with another fuel truck driver on the graveyard shift, and they were watching television between refueling calls. On the TV screen was a beautiful, well-endowed person doing a strip tease to seductive music. My husband's coworker was really enjoying the view until the dancer removed their breast covering and the breasts went with it. The dancer was a man, and the coworker was embarrassed and disgusted. He told my husband he better not tell anyone how much he had been enjoying the show.

Our next base was in Rantoul, Illinois, 1977. Our son was four at the time. He loved playing outside and would kick a soccer ball back and forth for hours. It was summer, and the grass was pretty brown and crunchy. He came in and asked for a jar to catch bugs in. I was raised in a rural area by depression-era parents. We saved good jars. So I gave him one. He came in later with it holding about ten grasshoppers he had caught. We had a lesson in bugs need air and punched some holes in the top. Later, after the kids had gone to bed, my husband and I could hear this odd clicking sound. We went in search of the noise and found the jar of bugs. I had pity on them and released them outside. My son was not happy in the morning when he found his jar empty. I told him he could go catch some more.

My other essay in this book explains my daughter's health problems when she was born. She had her surgery in Rantoul, and we learned that carrying a catheter bag was much like walking a puppy on a leash. She was 2 1/2 and just as cute as any puppy.

And then, we moved to Tacoma, Washington, where my husband was stationed at McChord Air Force Base when it was still an individual military installation. In the fall of 1978, when Tacoma was up to seventy degrees during the day, we took a day trip up the Mt. Ranier visitors center where it was about fifty degrees if you were lucky, and there was snow along the mountain roadways. After lunch, our son, just five years old, said he had to go to the bathroom, so he and Dad went off to the men's room. It took them an extra long time before they returned. My husband said, "I am never taking him to the bathroom again. He handed me his coat, went into a stall to sit down, then proceeded to sing some song and yell, 'Dad, you still out there?' before he repeated it. Everyone was laughing at me." I thought it was rather funny myself.

Our marriage ended soon after that, and I was no longer an Air Force wife.

Starter Home

By Lee Nelson

THE EX AND I WERE BOTH ACTIVE DUTY MILITARY, and we bought our first home in 1994 where we were stationed in Minot, North Dakota. Like all first homes, it was modest, but affordable, and when I say affordable, brace yourself: $39,999. Yes, you read that correctly: $39,999. Swear to god. Talk about things nobody will ever see again. $39,999 for a single home in America is probably dead as the dodo. It was a damn fortunate figure even in 1994, and I had the almighty location to thank for it. Property was cheap in North Dakota, so cheap that North Dakota probably loaned itself to the cheap property extreme that creates the national average. Property in Minot was the cheapest I'd ever known values to exist. We were just kids, and what did we know about a damn thing, but we knew a cheap mortgage when we found one.

We'd found this humble abode in the small town of Surrey, just a little east of Minot. It was twenty five minutes from base, peaceful, and $39,999! I never want to live in that house again, but I get weepy just thinking about how cost effective it was. I remember our mortgage, sure as the day Chapman shot Lennon: $405.75/month! You're drooling. So am I. You can't replace a front door for $405.75, and $39,999 is the sticker price on lots of cars these days.

And back to the location. Not only was this one toilet, two bedroom, dining room free, garage free, square footage free little bungalow planted in Surrey, but it was the house on the fringe that met the highway for the rest of the neighborhood. Our only insulation from the highway was a rather popular and prosperous steakhouse that everyone told us over and over we needed to check out. It wasn't particularly close to our house either. The one thing that tiny doghouse had going for it, other than affordability, was the size of the lot it was built on. The land was impressive. It was loud and open to anyone's view with no chance of privacy for thirty miles, but the lot was quite large. This I learned the first time I mowed the bastard. Anyway, the steakhouse wasn't exactly in our armpit. We didn't think living next to a prosperous steakhouse was going to be a bad thing, would you? What could go wrong?

Well, the part where they're not so prosperous of a steakhouse, that can go wrong. I did manage to get there once before it closed, not with my ex-wife, but with a bunch of my single friends. After all, it had a bar, lots of steaks, and it was near other bars. The food and the service proved lacking for a "prosperous steakhouse," and the dining room was more like a warehouse with industrial carpet and a few candles. I remember the bartender popped the tops off our beers with the old edge-of-the-bar maneuver. It's not every day you find a bartender with no bottle opener. Pretty classy. After that display of desperation, I did the bottle opening for the rest of the evening. I always kept a bottle opener attached to my key chain. Yeah, I'm classy too.

We got drunker than any collection of kids should ever get drunk in a "restaurant." There weren't enough bad steaks in the world to sober us up, and god damn were we loud. Ever play that game where you knock the top of your buddy's beer with the bottom of your own beer, and if your buddy isn't fast enough, his beer bottle sprays the entire world with a geyser full of Schlitz? We were doing that. It didn't go unnoticed. In fact, we sprayed a whole dining party. They were not impressed. Yeah, I have no idea how we didn't get kicked out of that "steakhouse," and then we went out drinking. This is the extent of my memory of that evening. You can imagine why.

The fine steakhouse closed its doors, and the vacant building was ok with me and the ex. Things got quieter, but the quiet was short lived, almost as short lived as our marriage. I won't get into those details. Let's just say the marriage survived longer than the house. We split up and went on hiatus, and later did a take-two back in base housing. However, in the midst of our hiatus, there I was with half the income, and suddenly $405.75 a month was all the damn money in the world. I was hardly enamored with my picket fence fantasy anymore. I needed to sell as fast as I could, when suddenly appeared a gaudy neon light flashing incessantly in my windows. Yeah, the strip joint that succeeded the steakhouse wasn't about to help matters whatsoever.

But my friends thought it was the greatest thing that ever happened. Why wouldn't they? Shit, it wasn't their house, and it was right next to a strip joint. Fuck yeah! Guess who wanted to stay at my house with no wife in it? Everyone with thumbs. It's one thing to enjoy your frat house years, but who wants to own the frat house? The good news is it was a one a done. Ironically, the depravity of that strip joint saved me from indefinitely sheltering about a half dozen John Belushis.

So, there we were, about five of us, pre-gaming in my doghouse the day of the one and done. We got all liquored up and made the trek across my lawn, opened the dark doors that reveal every strip joint ever, and entered a world most people would never talk about ever again. Let's just say it was rough, very rough and seedy, even for a strip joint. I couldn't tell you who did the zoning in Surrey, North Dakota, but suddenly there it was, a red light district devil's den across the highway from the church and the elementary school, and right next to my humble picket fence. Everyone in the joint was grizzled and older than their years. Naked, grizzled women smiled at us in coked up euphoria. It was hardly the Fort Lauderdale scene young men raised on 80's films expected to find, thinking it actually exists. My god, the inhumanity. We were begging the world for more clothes everywhere, and one of them invited herself to my lap. I was speechless, and suddenly felt flu symptoms. She finally got up, disappointed with my lack of financial gratitude, and I wanted to cross my lawn again and change my jeans.

Then there was the coked up waitress. You can spot that toothless coke-snort anywhere. The snorts never stop. It's a rather manic snort, and she manically took our beer orders. The beers showed up, and no bottles this time, nothing but cans of Keystone. We could have anything we wanted as long as it was Keystone. We cracked the beers ourselves. Why break precedent from the "steakhouse?" Honestly, we felt better cracking our own beers given the grime and the odors. I checked to make sure there was no idle cocaine on my beer, but I suppose I was always safe. That shit's expensive. Nobody wastes that stuff.

And that's when we all started noticing the Russians. Well…okay, we could've been profiling here, fair enough, but they sure looked the part. They were two impressively large, gritty and goony looking individuals at the end of the bar. They sported tight shirts that accented their bulging biceps, and they had eyes like rapiers, and black studded shit kickers. Their only expression was anger, and all of us seemed to be the only people worth looking at as they downed their iced glasses of bourbon, their personal bourbon, the bourbon we weren't getting. One of the two would lift his little glass to take a sip, and his fat fist would eclipse his entire face. It was odd. They made us feel really unwelcome. We were surprisingly behaving ourselves (probably because we were all nauseous and terrified), and you would've thought our money was good in any strip joint on the planet, but no, not in this cocaine-cowboy club. Not tonight. We dropped a few bucks on the table for our waitress's coke habit, and out the door we went. We went to other bars, of course. Then we crawled across my lawn well after last call and howled at the moon.

My friends had a ball tooling me and my tanking property value.

"Hey Lee, maybe Boris will let you borrow a cup of sugar."

"Hey Lee, maybe The Brady Bunch will mow your lawn tomorrow."

"Hey, Lee, wanna whack the town assessor? I know a guy."

I found myself in a real estate office desperate for an agent, any agent. Help! Help! I felt like I was looking for a cop. For god sake, somebody do something! Help me! And they helped me. I've never met a real estate agent that wouldn't close on their own mother, but she looked at the Bada Bing next to my house and didn't exactly reek of optimism.

Then one day came the most peculiar event. I was cleaning teeth (my military profession, as the military arbitrarily decreed) when my first sergeant tapped me on my soldier. My presence was requested in his office. That's never a good feeling. What the hell did I do? I finished with my patient, and to his office I went. The squadron captain had made the appointment too. Ho boy. This wasn't good at all, but I was incredulous. What the living hell is going on?

"You own a house in Surrey, is that correct," asked my captain.

"Yes sir," I replied.

"Hang on, airman. We're gonna make a phone call."

Dear Jesus, the suspense. I stood at ease as the captain dialed the phone. Lots of numbers. Collect to Europe? He handed me the phone.

"Is this Lee Nelson?"

"Yes sir, speaking."

"Lee, this is Agent so and so of the FBI..."

The FBI! The freaking FBI? The FBI and my house? This house is gonna be the end of me! Does Boris know this?

And that was actually a good question. No, Boris and Dmitri had no idea. At least that was the plan. The FBI was requesting use of my home for surveillance of the strip joint next door. They would even pay me! Well, hot damn, finally a little good fortune! I didn't even ask what they pay. I just told them to have at it and moved into a dorm room with one of my buddies and slept on the couch. I was done with Surrey, North Dakota. That sleepy little town was showing its warts. I was living in a B-film with drug lords hiding in plain sight. You couldn't write this shit. By the way, that's exactly what the FBI presence was all about. They wanted to stake out the strip joint for interstate drug trafficking.

I felt like I had to tell my real estate agent. I hoped I wasn't blowing the FBI's cover. They didn't exactly give me further instructions. She laughed. What's next? A devastating hailstorm? As a matter of fact, there was.

A hailstorm the likes nobody had ever experienced made its way through Minot and the surrounding Ward County one unassuming early morning. Some of the hail was apparently the size of bowling balls. Most others were soft ball sized. Can you imagine bowling balls falling from the sky? I never did personally see any, but the devastation was everywhere. Car dealerships with fleets of destroyed vehicles. Neighborhoods far and wide with decimated siding and broken windows. Downed trees.

Yes, there are some trees in North Dakota, believe it or not, and the hailstorm didn't improve their average.

Mobile homes crushed like beer cans in a drunken squeeze. Neighbors checking on neighbors. Is everyone ok? I didn't check on Boris and Dmitri. I'd decided they were resourceful types. I'm pretty certain they never needed me. It was very bad, that storm, but amazingly, it treated me very well.

The insurance companies rolled in, and all they wanted to do was pay everyone off and run for the hills. My home damage was luckily very minimal. I had splotches in the stain of my cedar shakes from some modest hail strikes. All they needed was a fresh coat of stain. I went on the roof and the roof was sound. The adjuster, however, did not do that, and he looked at me and said I needed a new roof and a new paint job. He wrote me a check for a lot of money, not enough to replace the roof, but a lot of money, and waited for me to grab it. I grabbed it like it was gonna run away, and that's exactly what he did. I grabbed that check and the man vaporized. Then I got the house re-stained, and there it was, ready for the market again. My profit was $1700, a sizeable amount for a broke, house poor airman.

The FBI made a lot of bold statements, but they never showed up. I never saw a dime from those guys. The realtor called me one day and said she had an interested couple. Around the same time, Boris and Dmitri closed shop. No headlines were ever made. It was all very casual. Another business just closed its doors. I wonder if they smelled a rat. Maybe the realtor tipped them off. I have to admit she sold my house no problem. The buyer took all closing costs. I broke even, and I was happier than a pig in shit. You could hear my sigh of relief clear across Ward County.

Everything, all of it, all the stress triggers of my starter home just disappeared in short order. It was like the circus finished its last show and left town. That was that. Peace out.

As I endured all this lunacy, my ex and I strangely drew closer. As I said earlier, this was all, at the time, a hiatus. Things were getting better and the future looked bright.

I might have even asked myself what could go wrong.

LD's Value

By John P. Bland

AS I AGE, I find myself sitting at my computer and looking online for old armed forces comrades. Recently, I found the obituary of one of my former commanding officers (CO), and what I read about the man astounded me. I had not known much of what was written. Like a magnificent iceberg, there was much more below the surface of the man than met the eye.

Commander LD Nagel, US Navy, was a quiet, almost shy Texan born in Round Top in 1931. While pursuing an engineering degree at the University of Texas, he was drafted for the Korean War. Instead of being inducted into the Army, he joined the Navy as an enlisted communications specialist and received a rare appointment from the Secretary of the Navy to the US Naval Academy. Graduating in 1956, he chose Naval Aviation as his career path. During his 26-year career, he flew the single-seat, single-engine A-4 Skyhawk attack airplane from the decks of seven aircraft carriers and became a highly decorated veteran with 194 Vietnam combat missions under his belt. Besides his BS degree from the Naval Academy, he also earned a second BS degree in Aeronautical Engineering and an MS degree in Aeronautical Electronics. He had been assigned duties in weapons systems design, testing, and procurement. Clearly, this man was a Naval Officer of high value.

During 1972-75, I served a Navy exchange tour, assigned as a flight instructor in Training Squadron Twenty-One (VT-21) at Naval Air Station, Kingsville, Texas. I checked into VT-21 in the fall, fresh from a combat tour of over 100 missions in Viet Nam. My welcome aboard was cordial and pleasant, but I didn't meet Cdr. Nagel on the first day because he had some more pressing duties. Activities around the squadron were intense because the end of the fiscal year was only two weeks away and pilot training quotas had yet to be met. I met the CO about two days later in his office in the squadron hangar.

Cdr. Nagel had the usual squadron plaques, certificates, pictures, and airplane models in his office. Instead of a container of jellybeans, M & M candies, or even a box of cigars on his desk, he had a large jar of jalapeños, into which he occasionally dipped, pulled out a few, and popped them into his mouth. We chatted about what he expected of an instructor, what additional duties I might like to have, and family matters. While I didn't learn much about my new boss, I left feeling LD was OK.

In talking with my fellow instructors about the CO over the ensuing weeks, I learned that LD was the CO's real first name on his birth certificate. We supposed that his parents, Lydia and Don Nagel, couldn't or didn't decide on a name and had combined the first initials of their names for his. His wife Lucy was a former Naval officer herself, and they had two grown sons. I would learn more about the CO from other squadron mates; as Paul Harvey used to say, "Now for the rest of the story."

On one very early summer morning, young Lieutenant (Junior Grade) Nagel was onboard the USS *Independence* cruising the balmy Mediterranean Sea and preparing for a routine patrol mission in an A-4. It would be a nighttime launch from the carrier and a daylight recovery with about two hours elapsed mission time. Donning his flight gear and grabbing his helmet, kneeboard, and flashlight, LD made his way from the squadron ready room to the flight deck where the airplanes were lined up, prepared, and waiting. He was met by a Plane Captain, an enlisted sailor whose duty was to ready the airplane for flight and assist the pilot in getting ready to start up the jet and taxi forward for a catapult launch. LD gave his gear to the Plane Captain, who climbed the aircraft boarding ladder and began to stow LD's equipment in the cockpit.

Meanwhile LD, flashlight in hand, began his preflight walkaround inspection of the airplane, which was parked on the flight deck's port edge. The tail section of the airplane was hanging over the deck edge, and as LD inspected the airplane's vertical tail with his flashlight, he failed to watch where he was stepping, tripped on a tiedown chain securing the A-4 to the deck, and fell overboard, sixty feet into the sea.

The Plane Captain, who had on his protective hearing earphones, didn't hear anything, and, because he was stowing LD's gear in the cockpit, didn't see anything, either. No one else on the flight deck saw or heard anything unusual. LD essentially disappeared. Sixty feet below bobbing in the sea and floating with the aid of his life vest, LD shouted and waved his flashlight to no avail.

Independence steamed onward, trailed by a Soviet "tattletale," an intelligence-gathering vessel. With sophisticated sensors and communication equipment and manned by capable crews, these ships disguised as fishing trawlers spied on naval activities and reported everything to Moscow. While not fast enough to keep up with most warships, they tended to congregate around aircraft carriers conducting air operations.

Fortunately for LD, the crew of the Soviet ship spotted him in the dark water. The vessel hove to, hauled up the soaked pilot, and contacted *Independence*. The captains of the two countries' combatants came to an agreement on the price of the return of Lt. (JG) Nagel. He was traded for five gallons of strawberry ice cream, his evident value at the time.

Made in Korea

By Tim Hansen

RVWG / Oct.2, 2020

ON MY FIRST OVERSEAS TOUR in the Republic of Korea, I soon realized that most of the troops loved to head into the ville after duty hours. The main street right outside the gate of Camp Pelham was lined with clubs, bars and cheap souvenir shops.

Now don't get me wrong about being anti-social or a teetotaling, regulation-quoting zealot of a lieutenant. Au contraire. I enjoyed sipping Irish whiskey and making small talk at the bar with others ... but not every night. As the Bard once wrote, "There is more to heaven and earth than can be imagined in any philosophy."

The sociology of the club scene was interesting because each ethnic group had claimed a particular club as its own. The local economy seemed to thrive with establishments such as the Cream Bar, Jo-Jo Club, and Cadillac City. How could that be? The Puerto Ricans had their club, the blacks had a club, and even the whites had one. We were one by day and many by night. It rubbed on my notion of equality, but then I realized that everyone was fine with this as the natural social order.

Not caring for soju and its effects of having one-too-many, I heard of a well-regarded master sergeant on post who lost count of how much he drank and of his running through the ville stark naked. Soju can make a man do some very unusual things and who needs to live with that kind of embarrassment? I thought.

Instead of drinking, I found a good tailor down in Tong Du Chon, known for making well-cut mess dress blues and other dress uniforms. I was so impressed with my dress blues that I ordered a set of white mess dress. Not thinking of any other commitments beyond my one to the Army, I wore my white mess three years later when I married my wife.

I soon started clipping ads for suits, jackets and coats out of *Esquire, GQ,* and *Maxim*. I had fun with it and started to make my own designs. On my weekly runs to Camp Casey to turn in signature sheets and cash from the dining facility, I would walk across the street while my driver was on his lunch hour and visit my tailor.

"Anyeonghasyeo, Ajusee!" I would say to him upon entering his shop and bowing courteously to him before transacting business. He was always very calm and gracious to me. I then would pull out my drawing and magazine clips and give him my next order.

"Now, Mr. Kim," I would say, "I want the cowl neck on this check to be just the same size as the one Bogart wore in Casablanca and here is my drawing."

I always thought Kim enjoyed my sketches, instructions, and magazine clippings. It was almost like throwing down the gauntlet and Kim savoring the moment of rising to the occasion to master the challenge.

Kim picked up on my sense of style. He nodded his head enthusiastically when I came in with my next order whether a double-breasted jacket with side vents like the one Carey Grant wore in *His Girl Friday* or the grey suit like Gregory Peck wore in *The Man in the Grey Flannel Suit*.

And, whenever I tried on my next suit, Kim would nod his head at me approvingly and say in a hushed tone, "Finest quality." And, indeed, it was the finest quality at a very good price for a young first lieutenant.

In time, I learned from Kim that he studied fashion and design in Paris. Paris! Why is Kim here in this dusty little town across the street from Camp Casey?

"Family," Kim would say simply. "I have taught my son and will teach my grandchildren."

Perhaps my most ambitious project for Kim was my bomber jacket. I saw one in a Banana Republic catalog that caught my eye and displaced me in time to sitting in the left seat of a B-17 flying at

25,000 feet over Germany. I liked the look and it connected me to the Allied effort in World War II.

When I tried it on weeks later, Kim gave me his reassuring nod of approval. It fit and like an artist, his grin conveyed that he had made something noteworthy.

"Kamsahamnida, ajusee," I said gratefully and bowed as I offered him my payment.

Just before I left, I had read a New York Times interview of Tom Wolfe. His signature sartorial statement was a white linen suit. Linen was a good material that held up to travel and hot weather. It also had a look of elegance to it. My tour in the Land of the Morning Calm had sneaked up on me and I ran out of time and I did not want to extend for another six months since I just landed my next assignment with 10th Special Forces Group at Fort Devens, Massachusetts.

Sadly, I never stepped into Mr. Kim's shop again.

Dealing with Mr. Kim opened my mind to design, cut, and fit. I started reading a little more deeply on how to look good and convey a sense of good taste. Even in this age of grunge wear brought on by the Silicon Valley culture, I try to find ways that express my personality by what I wear and it must be ... "finest quality."

Heterogeneous Recruits at Basic Training

By John Steele

I WENT TO BASIC TRAINING at Fort Leonard Wood, Missouri in early January 1986. That first day at the reception station, it was maybe thirty degrees, and there was maybe an inch of snow on the ground. Being from Rochester New York, no big deal. But people were totally freaking out like they never saw snow before. There is a reason for that. Many of them had not. I also made a friend that day, some guy from Alaska. To help freak everybody else out more, he and I took our shoes off and walked around outside a little bit in the snow. It worked, some people lost their minds, thinking he and I had lost our minds.

Funny side note on the guy from Alaska. The next day at the reception station, they took our clothes, gave us all matching uniforms, and shaved our heads. I could no longer find the dude from Alaska anymore. I couldn't recognize him. Everyone looked the same. They then sent everybody on their way to their own respective basic training companies, and I never saw him again.

All of us guys going through basic, we were an eclectic bunch. There was a guy from Philadelphia named Joseph. Joseph was his last name. No idea what his first name was. I didn't know anyone's first name. All I knew were people's last names, and where they were from. Anyway, Joseph had a full scholarship to play division one college basketball, but he got busted for possessing a certain amount of pot. He was found guilty, and was told it was either jail time, or join the military. He was forced to join the Army because of drugs. It's funny, because a few short years after that, people were denied military service for recreational use of those same drugs.

Joseph wasn't all that scary. We had another guy in our squad, Taylor, who was scary. Know the prison movies where they have some big imposing guy who never smiles and does nothing but lift weights all the time. That was Taylor. He always looked pissed off with a scowl on his face. He rarely spoke, and when he did, it was vulgar. Taylor was from California. He liked me. Let me change that. He didn't openly dislike me.

Now for a contrast, we had a guy from Utah. He was a Mormon. No idea what his name was, just the Mormon. We were not very politically correct back then, and it was a long time ago, my memory isn't what it used to be. The Mormon cried a lot, but that wasn't the biggest thing that made him so memorable. The Army implemented this rule where drill sergeants were not supposed to swear at us. One drill sergeant in particular had a hard time abiding by that rule, and the Mormon filed multiple complaints against him. Drill sergeants were also prohibited from physically assaulting us. But multiple times the Mormon had his fingers accidentally stepped on while he was doing push-ups. He was frequently singled out to do push-ups, and all of the drill sergeants seemed to have had difficulty avoiding his fingers with their boots. Additionally, anytime a volunteer was needed, you know who the drill sergeants picked.

Drill sergeants could be dicks. We had one guy who decided he was going to do something about it. Unfortunately, I don't remember his name, or where he was from. All I remember is he was Italian. We were close to the end of our second week of basic rifle marksmanship. Everyday after training as we left the range, we would have to clear our weapons and the drill sergeants would do a shakedown. We'd have to empty our pockets and line up, each of us professing individually that we had no brass or ammunition and we would be patted down. Then we would march back to the cantonment area and turn our weapons into the arms room. The arms room was a separate concrete building, maybe fifteen feet by twenty feet in size, with a door on each end. You walk in one door, up to the counter, call off your weapon number to turn it in to one of the drill sergeants, and exit out the other door. Each platoon would go through by squad. As first platoon was going through, out of nowhere, a loud shot echoed off the concrete walls. All of the rest of the drill sergeants swarmed like bees into the building. Evidently this recruit had managed to keep one round through the shakedown, and at some point chambered it. As he went through the arms room, he took aim and fired his lone shot at the one drill sergeant in the building. Even though the entire building was only fifteen by twenty feet in size, and this recruit had spent the past two weeks doing nothing but basic rifle marksmanship, he missed. Army drill sergeants must be lousy teachers. That recruit was never seen by us again. I am sure he ended up at Fort Leavenworth, home of the Army's maximum security prison.

We actually had a second guy from Utah. This one was named Synderguard. He was a Mormon too, but he wasn't a very good Mormon. He was as normal and average as they come. He blended in as well as anyone. Average size, average weight, average intelligence, average physical endurance, as non-descript as possible. The exception was when we had our one weekend pass. Synderguard drank like a fish. Thus, my reason for saying he wasn't a very good Mormon.

We had another guy from Pennsylvania too, Bishop. Tall and lanky, he was smart as a whip. He was the brightest out of the group, by far the most intellectual. Also the least athletic. He really struggled with the physical side of training. He was bespectacled, wearing what we commonly called BCG's, for birth control glasses. BCG's were the Army issued eyeglasses with the thick, black frames. If anyone remembers Drew Carey from his early days as a comic or from his Drew Carey Show, he wore BCG's.

There was a guy from outside of Richmond, Virginia named Moore. Couldn't understand a damn thing he said. He was nice enough. A country boy through and through, he chewed tobacco continuously, and had the worst looking teeth I ever saw.

Next up, we have a guy from American Samoa, Emmett Soiafua. I did know someone's first name. He was probably my best friend in basic, and I am sure that I am spelling his name wrong. Soiafua was built like a redwood tree. Shortly after we got there, we had an Army Physical Fitness Test, and he maxed all three events beating all 120 of us in the company. The first week of basic, we would get dropped for twenty or thirty push-ups, Soiafua would do them one handed. He never looked tired, and he was always smiling. Smiling so much that every time our First Sergeant saw him, he would drop him for a hundred push-ups. Soiafua would gladly do them, calling them out as he did them. After the hundred, he would ask if he could do more. Nothing bothered him. Soiafua is another guy who had an athletic scholarship for college. He turned his down. College athletics was a business, and he didn't want the commitment that was required. He came from a big family. His goal was to stay in Samoa with his family and marry his high school sweetheart. That's all he wanted, a simple life.

The biggest aberration for me at basic training, aside from the guy who attempted to assassinate our drill sergeant at close range and missed, had to be my battle buddy, Smith. He was from Galveston, Texas. He was only seventeen years old. He was the youngest guy in our squad, by far the most immature, and the only one who was actually married. The crazy part was his bride. She was only fourteen years old. He matter-of-factly told me that obviously they needed to get parental consent for her to marry him. Her parents were divorced. One parent was against it, so of course the other parent agreed to it just to piss off the first parent. So the one parent agreed to sign the papers not because they approved of their daughter marrying Smith. No, just to upset their former spouse. Smitty dropped out of his senior year of high school to join the Army. He wanted a job so that his bride could stay in school and finish her freshman year of high school.

Me? I was the elder statesman. I was no longer a teenager, and had completed over a semester's worth of college courses. Being from Rochester, NY, everyone assumed I was fresh off the streets of New York City. That is until they met me and got to know me. I was a city boy, but it was pretty obvious that I wasn't from the big city. Moore, being from around Richmond, couldn't fathom that I lived in New York state, but that it was over 350 miles from New York City, and that I had never been there. He and his father drove there once, and it was about 350 miles away. They drove through Virginia, D.C., Maryland, Delaware, and New Jersey to get to New York City. How in the hell could Richmond be closer to New York City than Rochester when Rochester was in the same state?

We were a rag-tag bunch, coming from all over the United States. We all had different backgrounds, and different experiences. My America was very different from Smitty's, which was different from Soiafua's, or Synderguard's, or Moore's, or Taylor's, or Bishop's, or Joseph's, or the Mormon, or the dude from Alaska who I never saw again. But we were all Americans, coming together to serve our country.

Beware of the Seabat

By John Michael (Mike) Ryan

IN 1970 I HAD BEEN ABOARD the Springfield for eighteen months and I had heard my share of stories about on-shore, hometown, on the job experiences plus fantasies and gags. One gag in particular comes to mind.

To set the scene, the 6th Fleet was on patrol somewhere in the east end of the Mediterranean Sea. I was in my print shop looking down over the deck and at a calm sea. It was a sunny weekend day. Twenty or so crew members were milling around either doing their job or enjoying some free time. Some must have known about the gag, but kept silent until the end. On the deck was a large, upside down cardboard box.

The 1st Class Boatswain Mate (E-6) had a mission that day involving the box. It was to introduce some unsuspecting sailor to a mythical sea creature known as the Seabat. With help from fellow instigators, who searched the ship for victims, he waited for someone to show up to see the captured Seabat. When one did show up, he was told to carefully lift the box and to take a look. When he bent over to do so, the boatswain mate grabbed a huge flat-faced broom and whacked him on his buttocks.

The witnesses laughed loudly to the consternation of the victim. The beauty of the scheme was that the victim kept quiet rather than admit that he had been tricked. That paved way for the next victim. That day, from my table top at the hatch, I saw five unsuspecting souls try to take a peek at an invisible Seabat. I was never a victim during my time on board.

Second Amendment Adventures-4

The Light at the End of the Gun Barrel.

Chapter 4

By Vaughn Stelzenmuller

OH COME, COME. Nothing that dramatic. But it *is* good for readers to know that all firearms in the park are now unloaded, locked, and open. Which means that an inspector could look into the chamber (that place where all the bullets sit, waiting to be trigger-pulled and shot out the barrel) and see shiny light coming from the barrel's muzzle end. This means that no one could shoot, nor be shot. Miracle!* But how did we get here?

In the most UN-original military idea going back to *Umpteen* Hundred BC/BCE, I go around to every "campfire," looking the small groups in the eyes respectfully and firmly, speaking personally to as many as time allows, inspecting the safe state of the arms on the ground (hopefully), and mentally selecting subordinates. All at the same time. *David Niven* can continue handling that well; just try to look as collected and handsome as that good actor. Without moustache.

Assume that somehow I was able to sort all this through. There will certainly be persons in that field who have good police or military history to pull off a rough organization, battalion size, if respectfully chosen by *David Niven* by the early afternoon.

Also, if McDonalds™ and Seven-Eleven™ come through with some kind of meals promised by the City Council and Councilman Richard "Jake" Johnson. An "Army travels on it stomach," supposedly said by Napolean, but a civilian sort-of-army doesn't even stay in one place through the afternoon without being fed. And watered, non-alcoholically.

Some of the choices for subordinates will not be great, but most will. The key, of course, will be the group's *belief* that their new chosen superiors *are* superior in experience, and will always make the right choices on their behalf. Most of the time, this really works. Along with "visiting each campfire" in this odd situation, the new "sergeants" are left with their small groups to form them up, get them pointed in the same mental direction, and ensuring that all firearms are unloaded before any drilling. And not really in that order.

That is how we got to this point. The day ends more or less successfully, though I let it to some other historian to tell "The rest of the story." Mr. Paul Harvey is unfortunately deceased, but some smaller, less-expensive budding talent will surely fill the gap. And help me not go to jail after all.

*Speaking of miracles, no one died during the unload-the-gun period just ended. No ambulances, no cable news spectaculars, and no handcuffs for me.

What did all the Second Amendment bustling around accomplish? As an unlikely true-event itself, not much. What does matter is what it could *show*. That amendment to our Constitution, one part of the ten-part Bill of Rights, points out the potential defense that a free citizenry of free republic could form in the face of external threats. Clumsy and dangerous though it would be, its possibility displays its worth. That possibility makes those "threats" not want to take a chance in the first place. Particularly if some elder *Robert-Redford* type retired officer tries his hand at pulling off this unpredictably effective organizing.

and
We laughed

LAST CHAPTER: UNCHAINED MILITARY HUMOR

Some military humor defies classifying, and perhaps these final writing selections should be classified:

"Top NON-Secret"!

Hot Dogs

By Lee Nelson

Yes
they are bologna
but for all the
grinded
squished
pressed
and compacted
gross entrails
tongues
hoofs
snouts
and solid organs
that birth
the sheet meat
and well formed
defecations
via tried and true
ancient technologies
of simple will
and weight
over surrendered
and grizzly
executed
carcasses

we're not talking
birds of a feather here.
Even as the birds
slap

your sliding glass window
at the speed
of absolute cluelessness
not even close

and for fully
unnecessary transparency
(no pun intended)
I once ate one of those birds
so I would know.

Even the best bologna
is tasty sustenance.
Nothing more.

Just the right
hot dog
can be the most
gloriously
hedonistic
decadence
of shameless
and heartless
culinary invention
straight from the 'taint
of the ugliest god.

All you need is
copious onions
a meager mustard
and an even more
meager
Wonderbread roll
and the fullest flavors

of sinful Hell
bear their undignified elations
to your tongue and teeth.

You can go to Hell
for a lot of reasons.
Eating a hot dog
isn't among them
so you may as well
eat Hell's food
enjoy life
and hopefully
go to Heaven.

A hot dog
can fill this bill.
Bologna
doesn't stand a chance.

Being Abandoned on Thanksgiving

By John Steele

The most memorable holiday for me while I was in the service is Thanksgiving 2011. The unit returned from Iraq around Veterans' Day, and our de-mobilization station was at Fort Lewis, WA. We spent a week or so going through the demob process, out-briefings, physicals, turning in weapons and various equipment. After about a week, soldiers from my unit started going home by piecemeal, but if there were any medical issues, you were kept longer for further testing. I had been having issues with my foot, and after an MRI, they had ascertained that I had three torn ligaments and two torn tendons. Long story short, I needed reconstructive surgery on my foot, and until then I had to wear a boot and I was told to stay off of my feet as much as possible. The Army was going to put me into the Warrior Transition Unit or WTU on base. When would that be? They didn't know, but be ready.

Well, on Wednesday, November 23rd, at about 1500 hours, I was told to get my stuff together, someone would be over to grab me and take me over to the WTU, but I had to be there by 1600, as they were closing up shop. I was ready to go by the time they came to get me, and we got over to the WTU just after 1600. The WTU was already having its closing formation, so I heard the tail end of the safety briefing, a variation of "don't do anything stupid." Then I was told that we were dismissed until the Monday morning formation at 0600. Holiday weekend, all the full-time people were ready for a long weekend. So I was quickly introduced to who would be my squad leader. He shook my hand, handed me a key, and said, "I'll see you Monday morning for in-processing." "Wait, what is the key to?" "Duh, your room." "Where?" He pointed to a huge building, and said the room number was on the key. CQ (Charge of Quarters) could answer any and all of the basic questions. And he left. I was abandoned.

The building, the barracks and my new home, was more like four buildings connected together with at least a hundred rooms. So me and my bad foot in a boot had to drag all of my belongings from the past year in Iraq that were stuffed into or tied to my ruck sack and duffle bag up to the building. I found the CQ desk. Empty. The sign said that with it being a holiday weekend, CQ was gone 'til Monday morning. Happy Thanksgiving. I really was abandoned.

I still had to find my room. At least there were rooms, not an open bay. My room number started with a 4, so I figured out my room was up on the fourth floor, and luckily there was an elevator. My room had a bed, but no linen. I had no vehicle, so I couldn't run off to any store. My foot was in a walking boot, and I was told to stay off my feet. And I didn't even know where the chow hall was.

I went back down to the common area, but it was like a ghost town. There were a couple of junior enlisted who were much more interested in the video game they were playing than helping some lost master sergeant. They did tell me that the chow hall was in the basement of the hospital. It was about a quarter mile away. So me and my walking boot headed for the hospital.

Thankfully, I did find a familiar face in the basement of the hospital. A member of the unit I was mobilized with for nearly eleven months was there, Captain Tedesco. She had been put into the WTU the day before, so she was an expert. She was there because of her knee, and she was told to stay off her feet too. There was a shuttle that was supposed to go from the barracks to the hospital once an hour, but we learned quickly that it was shaky at best. We also learned that the chow hall in the hospital would be closed tomorrow for Thanksgiving. The shuttle would not deviate from its route to take us to a different chow hall. The only chow hall we knew of was on the other side of the post, on the other side of Interstate 5, basically a whole different post. We still had no car. So we had to find a different chow hall, and find a way to get there.

The next day around noon, Thanksgiving 2011, Captain Tedesco, in her mid-fifties with her bad knee, and I, a forty-eight-year-old master sergeant with my foot in a boot, hobbled in uniform, neither of us had any civilian clothes, down towards the center of post on the side of the road, there were no sidewalks, hitchhiking, trying to find where the chow halls were on this side of the post.

Someone did pick us up, and we did find one. It was a big chow hall, and they were pulling out all the stops for their Thanksgiving meal. The post commander was greeting everyone upon entry, and a whole bunch of brass were serving the meals. As we both limped up, the commander saw our unit patches, and asked us what unit we were from, he didn't recognize the patch. "The 804th Medical Brigade."

"Out of where?"

"Devens, Massachusetts."

"What are you doing in Fort Lewis."

"We demobed from Iraq, and were just put into the WTU."

"Why were we at this particular chow hall?"

"It was the first one we found that was open."

"Why is the one at the hospital not open?"

"We don't know."

"So you drove around looking for chow halls?"

"No, actually, we don't have cars. We walked and hitched asking people where a chow hall was, and someone brought us here."

"You walked? And you're limping, and you came from the WTU?"

"Yeah." (Or probably more accurately, "Yes sir.").

"Well how are you getting back to the WTU?"

"Don't know. Walk? Hitch or beg for rides?"

"No, no, no, I am going to get you a ride back to your barracks."

He had lots more questions for us. They mainly centered around how and why we were in the predicament we were in. And we didn't have very many answers for him. But we had a great meal. Different, but great. The turkey wasn't deep fried like my family usually has. The turkey was kinda' dry, but the ham was good, and my family never did ham on Thanksgiving. They had a number of side dishes that I didn't associate with Thanksgiving either. They had mashed potatoes, but instead of sweet potatoes, they had macaroni and cheese. Who does that? I asked, and I guess it is a southern thing. I know, we were in Washington State, not somewhere down south. Maybe the person who ran the chow hall was southern. I don't know. But I love mac and cheese, so that worked out. And they had this string-bean casserole. It looked gross, so I skipped it. They had something called okra. I still don't know what it is, and I refuse to google it. It too looked gross, and I skipped that too. And they had something they called greens. It was greenish, I guess, it had unidentifiable chunks of non-green things in it too, so I skipped that also. For dessert, they had pumpkin pie, which I'm used to and like, but also pecan pie, and peanut butter pie. I stuck with the pumpkin pie.

Since we had no transportation for an evening meal and our chow hall was closed, we were encouraged to take a second meal home with us. More ham, mashed potatoes, mac and cheese, and pumpkin pie. And the commander arranged a ride back for us.

Captain Tedesco and I were friends in Iraq, but we grew closer when we both spent several months in the WTU out in Fort Lewis. And every year since 2011, on Thanksgiving, Captain Tedesco and I contact each other and reminisce about the being abandoned out there for Thanksgiving.

CHUCK WILLARD TRIBUTES

On hearing of Chuck's passing. May 14, 2024

Comments from members of the Rochester Veterans Writing Group

By Sue Spitulnik

We lost our friend and brother this morning at 2:15. He left word we were not to cry tears of sadness but to celebrate that our paths crossed. Damn, it hurts.

-Sue Spitulnik

Sue, sorry to hear of your loss. Tears and sadness help show he was loved.

-Abbie Cole (wife of Dave Cole)

Oh my gosh I'm so sorry to read this news. Chuck was truly a great man. I treasure the lampshade that he made me. I have it on a driftwood lamp that my dad made when they had no money back in 1953. Such a hero and friend. We miss you buddy.

-Steve McAlpin

God Bless Chuck - so many good memories.

I'll abide by Chuck's word to celebrate that our paths crossed, but that won't stop the tears.

-Joe Mele

We are the less without our friend and fellow writer, but he left us with his wonderful stories about flight, the Air Force, and the meaning of service to his nation, family, and fellow airmen. I still see him reading his stories and then patiently responding to our many questions and comments. His wisdom and dedication to this group added greater depth and meaning to all of our meetings. My heart is very heavy with this news, but I am thankful that our paths crossed and we had the time to get to know one another.

A final salute to you Chuck, and through the mercy of God, may you rest in peace.

-Tim Hansen

Oh my friends. I've not seen you for some time but have tried to keep up with the prompts. Dear, dear, Chuck. There is so much to this bigger than life man. I hear his voice and can read and feel the passion he put from his mind to his heart and to his hand as he shared a most extraordinary life. Sharing his life and stories, not to mention his compassion and quiet listening ...I feel as you do. I honor you Chuck. Fair Winds and Following Seas. Rest in Peace knowing we'll all be here holding you in our hearts.

-Suzanne Dianetti

I've been mentally digesting this all morning, and really, I just have no words. There's no words in the world. Chuck was just an incredible, accomplished man, a worldly man. It's been a privilege to share this writing passion with him, and he sure taught me a few valuable lessons in the short time I got to know him. I think of comparative men that I've known, men of great capability and accomplishment. Often times, arrogance and intimidation go hand in hand with these accolades, but never ever with Chuck. Not a chance. Chuck was a humble man with a big heart, and he was here to share and help. I'm thankful for my fortune. I'm thankful to have known him, and thankful for how he touched my life.

Even the longest lives are too short. May he Rest In Peace.

-Lee Nelson

Chuck was my editor/mentor on my book contributions. I did not know him as long as some of you old timers but I know patience, humbleness, love about all aspects of his fellow humans and COURAGE when I see it. His shadow has not become smaller, it remains ever large.

-Richard Terboss

Remembering Chuck Willard

By Tim Hansen

I remember Chuck started attending our monthly meetings at Writers and Books back in 2015. His well-dressed appearance struck me as the corporate type and his black-frame glasses made me think of an engineer. His quiet and unassuming nature ended when he would read his stories about flight and serving in the Air Force in the early sixties. While listening to him, I almost felt like I was sitting in the cockpit of a C-130 at 30,000 feet and feeling the crushing stress of making life-and-death decisions over a sudden loss in oil pressure or engine malfunction while in flight. He captured the beauty of flying as well as all of the critical actions for taking off and landing. His stories always taught us something new about aircraft and the Air Force.

Chuck quickly became a regular at our meetings and generously offered his help to other members for editing their writing or advising on technical accuracy of a military operation. He supported our Veterans Day celebrations and always read some very graphic battle scenes he had been working on.

As the years passed, we came to know Chuck through his stories. He grew up on a farm in central New York where everyone had to do chores and work hard just to survive. I think the hardscrabble existence during his early years toughened him physically and mentally, but as hard as those times were, he always spoke positively about them and with neither bitterness nor remorse. I think he said once that he was the first in his family to go to college and his time at Syracuse University opened up his mind to mechanical engineering and Air Force ROTC. Chuck served on active duty during the tense times of the Cold War between the U.S. and the Soviet Union. In his first duty assignment at

Charleston Air Force Base, South Carolina with his first wife Lynn, he starkly contrasted the charm of Southern living with the possibility of nuclear annihilation at any moment.

Two years in a row Chuck opened his beautiful home out in Rush to the group for its annual one-day retreat. He showed us his writing cabin that he modeled after Mark Twain's cabin in Elmira. I remember that warm, sunny day as we sat on his porch and shared drafts of our stories in various stages of completion. I felt a true sense of community with all of the members that day - thanks to Chuck's thoughtful generosity.

I recall Chuck writing prodigiously through the years. He even joined another writing group in Rochester to hear other writers and read his works to that group. When the RVWG started publishing its anthologies, he generously contributed several of his stories and played a big role in editing the stories of other writers.

I learned from others about his work in developing the laser printer at Xerox and his ascent into the upper ranks of management, but interestingly, he never talked much about it. And, I always wondered why. Did he work closely with Gary Starkweather who invented the laser printer at the Xerox Research Lab in Webster in 1969? What were those days like at Xerox? Chuck had so many intriguing aspects to him.

After Lynn died in 2019 from her fight with cancer, Chuck took a break from the group to grieve the loss of a wonderful woman. All of us in the group missed Chuck that year until one day he showed back up with stories to share and news of his engagement to Evelyn Farrell from Arizona. Though Chuck and Evelyn moved out to Arizona permanently, Chuck steadfastly kept in touch via the internet teleconferencing venue known as Zoom. His reports of 80-degree weather out on the golf course kept us warm during

Rochester's winter months. Amazingly, Chuck never slowed down. He completed his autobiography, played golf, kept up with his wood working, and bike rides with Evelyn. When Chuck shared the news of his cancer with us, we all felt confident he would overcome it and continue to write. News of his death surprised and stunned me. How could he be gone? He always had a young and healthy look. I am still grappling with his departure.

Though I wish you could have lived another decade, nevertheless, may you be at peace, my friend, and know how much I admired your storytelling and so appreciated your unwavering friendship.

Rochester Veterans' Writing Group at Writers & Books, Rochester, New York, in 2016.

Left to right, back row: Vaughn Stelzenmuller, Kurt Feuerherm, Tim Hansen, Janice Priester-Bradley, Chuck Willard; front row: Steve McAlpin with Reagan, Sue Spitulnik

By Lee Nelson

Dear Chuck,

All I can truly say about Chuck Willard, I said to him in a final email between us, and I copied and pasted it below to serve as my tribute to him.

I got to know Chuck slowly and sparsely over the course of about six years. Little as those interactions were, every one of them was impactful and significant to me, and quite honestly, to my life. As we all know, we can't pick our family, but we can pick our friends. Sometimes, the friends we choose remind us of the best family we have, so we choose them. Sometimes, not at all, so we choose them. Sometimes, when good fortune befalls us, our lives are introduced to others whose design and purpose (to us and them) is to enrich and fulfill us, and bring us some place greater and more enlightening. Sometimes, we even meet people who do as much for our souls.

I'm not a guy who trusts easily, and I'm not a guy who welcomes easily. Having said that, forget it. None of it matters. Chuck was simply a high caliber human being. He arrived graciously with sincerity, a great heart, and smarts and wisdom. He was genuine and good, and you would've been a fool not to embrace him, welcome him, befriend him, and learn from him. I'm fortunate to have known him, brief as our time together may have been. He bettered my life, and he made a soul connection on me, and I will miss him.

With that, what follows is our final communication. Rest in Peace, Chuck. You were a best. As I know the world, you were a best in this world.

* * *

Chuck, I often begin casual greetings and emails with "I hope all's well," and I do. I hope all's well with you. I hear this isn't the case, but I hope all's well with you. I hope you're in comfort, and I hope you have all you need. I know you're surrounded with deep love, the deepest love life ever knows. I know that much, and to be honest, this comforts me. This is the best comfort in the world, an assembly around you of some of the deepest loves of your life. It comforts me to know you have this comfort. For all your misfortune, it's good to know that a good man has all he needs, and you're a damn good man. We've known little of one another, but you know a damn good man when you meet one.

Chuck, you've impressed me every time I've shared your company. Honestly, you've impressed me and intimidated me, intimidated me in a good way. You're an accomplished man with honor and a good heart. In the little time that we've known each other, we've shared joys we have in common, and in case you're unaware, you've mentored me once or twice. It hasn't taken you much to accomplish this. All you've had to do is be you, be you doing what you love. That's all you've needed to do. You're a worldly man with a few years on me, and you've given me a lot of insight I'll never forget. This writing passion we share, it isn't easy. It isn't easy even with practice and skill and talent. It's simply a difficult act in and of itself, and you've had great courage in sharing your stories. I thank you for all you've shared, and I thank you for all your sharing has taught me.

It's been deeply enriching to know you, Chuck, and I'll leave you with that. Be in comfort and grace and the love of your family. Knowing they are there is to know all's well with you. Take care, my friend, and my best to you and your family.

Yours Sincerely,

Lee

* * * *

Thank you Lee.

You are one of the most articulate people I know. Your note is a fine example. It made me tear up, usually a pretty fair indicator. Anyway,I love you, man. Give my best to all.

-Chuck

By Joe Mele

God Bless You Chuck.

Your eloquent writing lofted us skyward with you on your Air Force Adventures, and brought us along with you on your life's well-lived journey. We'll forever fly with you on the hulking four-engine Hercules C130. We'll barrel roll, loop, and spin in the Tweety-bird T-37 training jet.

We'll miss your "elder statesman" presence at our RVWG meetings, and your mentorship and friendship.

You asked us to be grateful that our paths crossed, and to keep tears to a minimum. The gratitude part is easy. We'll work on the tears.

Love you Chuck !

Chuck Willard Remembrance

By Richard Terboss

Chuck was my reviewer and mentor on some of my content that is going into Book III. He urged me to get more personal and to think more than "just the facts, ma'am." It helped. When I met him, I don't think I had ever met someone with his outlook. Chuck faced danger, the unknown, love and life without blinking.

I only hope that I can mirror this behavior. I hear you Chuck. What more can I say?

Signing Off;

Thoughts on Chuck

By John Steele

Chuck was an amazing guy. I remember when I joined the Rochester Veterans' Writing Group, he was there that first day. We got some prompts, and were given thirty minutes to write. I didn't consider myself a writer, and in those thirty minutes, I filled maybe three quarters of a page with a rambling gibberish. Then we took turns reading what we wrote. When Chuck read his piece, it confirmed that I was no writer. His writing was so eloquent. It was fascinating and descriptive and it pulled you right in. He also wrote so much in those thirty minutes, it would have taken me hours just to get the volume of what he put out in that half hour.

I was embarrassed to read my meager little piece. But after I did, Chuck was the first person to comment, and it was positive feedback. I downplayed his compliment, but the more I tried to diminish what I wrote, the more affirming he was. He is the biggest reason I returned the next month. He welcomed me and made me feel like I had something important to say.

That was Chuck. He was one of the best, most polished writers, but he never looked down on anyone else's work. Over the years, he was always positive and encouraging. The more I got to know him, the more I liked him. He had a good sense of humor. He was someone who I looked forward to talking to each month. I was shocked to hear how old he was. He presented as someone much younger, both in his boyish grin and outlook, and the bounce in his step. He had a wide variety of experiences and knowledge, and he was willing to share his wisdom when asked.

Chuck and his lovely wife Evelyn moved out to Arizona. I love road trips and seeing the country, and the American southwest is one of the only corners of the country that I have yet to explore. I was planning to vacation in Arizona and New Mexico, and Chuck was happy to recommend his favorite spots for us to visit. I am sorry that I didn't do it sooner. I missed the opportunity to spend more time with my friend and enjoy his company. He is and will continue to be missed.

Thirty-Thousand Foot Runway

By Vaughn Stelzenmuller

You have always landed your aircraft, your entrusted property and priceless passengers on any runway man or God had thrust in front of you. Some of those runways have undoubtably loomed close, but you always pulled controls, and then pulled through some pretty dire setups.

"God may be your co-pilot" like that titled classic, but just as important, any one of us would be your volunteer *passenger* right behind you, at ease with you as our airborne guide. And constant all-around guide you are, from the day you walked in years ago with your farm-&-goat story, to the champion writing contributor of our group of today. Your leadership never seems to stop, and is always needed badly somewhere. You have proved that, I suspect, all your adult life—likely even much earlier.

Your runways may be thirty feet, or thirty thousand feet. You would get us all safely landed.

To My Friend, Chuck.

By Sue Spitulnik

Who is the tall man that was once an Air Force C-130 pilot
And still identifies that way?
Who is he that has to look down at me,
But has never looked down *on* me?
Who is he that didn't laugh when I asked to interview him
But instead offered to help with my writing project?
Who is he who instilled courage in me,
And who praised my creativity with a smile?
Who is the guy who shares his experiences,
Never with a haughty air?
Who is he who willingly teaches,
And happily contributes at any time?
Who is he who organized the Rochester Vets Writing Group essays,
And kept them all in a growing file?
And who expanded the file,
With so many of his own escapades from all over the world?
Who is he who dealt with loss and grief,
Then showed us how to move on,
Working at finding the perfect new partner
Driving thousands of miles to get to know and then claim her?
Who is he who shares his home,
Both in Honeoye Falls, NY, and Green Valley, AZ,

with family, friends, and fellow veterans,
Always with love, extra food, and a good bottle of wine?
Who is he that can turn wood to paper thin,
Draw his own book covers,
Co-pilot a plane to Israel
And led the team that developed the laser printer?
Who is he who has become my big brother,
Filling a need in my life for a positive male influence,
And treats me like I want and need to be
So I understand my own self-worth?
Who is he who has a wife that calls me family,
Who shares secrets like a loving sister,
Who opened her arms and heart to
A fellow home-front warrior?
He's my very special friend Charles F. Willard, Jr.
Who goes by Chuck, Skip, Dad, Uncle, and Brother.
He adds a great deal to my life
And Bob and I are blessed to know him.
Thank you for being you
And accepting the me that I am.
With much love and adoration
Big hugs, though I can't reach that high.
Your sister, Sue Spitulnik.

March 22, 2023

AUTHORS

Janice Bilohlavek

My name is Janice (Harper) Bilohlavek. I served in the Navy from April 1964 to Nov 1966 at Long Beach Naval Station. I worked in the Financial office as a Disbursing Clerk. I was married while there. I grew up in Grandville, Michigan (near Grand Rapids) and moved to Rochester, NY in 1967 as my husband's family lived there.

John Bland

John Bland is a military kid of a Marine Corps fighter pilot. As such, aviation is in his DNA. He recently voluntarily gave up flying after 55 years as a private pilot, a Marine Corps fighter/attack pilot, a commercial airline captain, and an aircraft charter pilot. He has landed an airplane in every state, on over half the continents, and on three aircraft carriers. Born in Texas and raised all over the place, he makes north Georgia his home. One of his favorite books is, I'm not kidding, *The Great Santini*.

Cindy Blair

AG1 Cynthia (Cindy) Blair grew up with a love of God and Country as the daughter of a Korean War Veteran in Upstate NY. She felt the call to duty after the fall of the Berlin Wall and in September 1989 signed up for the Navy on a delayed entry program and worked in the recruiting office.

May 1990 - left for basic training in Orlando, FL; Jan. 1991 - Aug. 1995 - weather observer at Belle Chase, LA

Aug. 1995 - Aug. 1998 - weather observer and assistant forecaster at Sigonella, Sicily, Italy; Aug. 1998 - Sep. 2002 - Meteorological Research Assistant at Fleet Numerical Meteorological and Oceanographic Center in Monterey, California. She continues to serve in her community in various volunteer and peer support roles.

Chris Cilento

I served from 1995 to 2021. My first five years in the Air Force, the rest in the Army. I have had several combat deployments that profoundly changed the way I both saw and interacted with the world and myself. I have seen some of the worst parts of humanity and left pieces of my soul behind. I became empty, hollow, and angry. It was only when I rediscovered my childhood love of art and writing that my life dramatically changed. I've found that art and writing have become powerful and transformative parts of my journey. They've helped me navigate the challenges of Post Traumatic Stress and chronic pain, while also opening new pathways to healing, growth, and self-discovery. For many long years, my world was shaped by the structure and intensity of military life. Now, through creative expression, I've begun to see and engage with life in new and inspiring ways. My artistic practices have become a source of strength, resilience, and renewal, helping me reconnect with the beauty and meaning that surrounds us all.

Dave Cole

1968 - 1969 served with 1st and 3rd Marine Division as infantryman and Squad Leader in both Divisions.

Village of Wayland Police Officer, NY 1975-1977

New York State Police 1977 - 1999

Steuben County NY Undersheriff 1999-2012

Steuben County NY Sheriff 2013 - 2017

Richard DeLisle

Dick was born in Buffalo, NY in April of 1940. He played baseball, basketball and track during his high school years. His basketball skills got him a scholarship to Alfred Tech.

Despite his academic ability, he soon found out that he wasn't ready for college yet. He dropped out and enlisted in the Army. It was one of the best decisions he ever made: he learned to give and take orders; to plan projects; to manage time; and to develop long lasting friendships.

When he got out of the Army he enrolled in college and made the Deans List every semester. At graduation, he was hired by Xerox and moved to Fairport, NY. He subsequently moved to Webster with his family. While in Rochester, Dick attended The University of Rochester and graduated in 1968. The GI Bill made that possible.

He left Xerox after five years and for the rest of his career he worked at Raytheon, Westinghouse and Figgie International (Scott Aviation). He retired as President of Scott Aviation in 1989.

Dick and his wife Elaine have spent their retirement years volunteering. They spent twenty-eight years with Habitat for Humanity doing electrical wiring. For the past four years they have been delivering Meals on Wheels five days a week.

They have eight grandchildren and going to their activities keeps them busy.

Dick proudly wears his Army Corp of Engineers cap.

Tim Hansen

Tim Hansen is an Army veteran and cofounder of the Rochester Veterans Writing Group. He has worked as a public affairs officer in Afghanistan, Kuwait, and the Pentagon, and currently works as an editor for U.S. Immigration and Customs Enforcement.

Dana Hill

U.S. Air Force 1972 to 1992. Dana retired an E7 from Biomedical Equipment Repair. During Operation Desert Shield she went to England to set up a 500-bed POW hospital. For better or worse, the hospital never got used. Dana was the first female in that field to make senior noncommissioned officer before retirement. She has often wondered, "Should I have stayed longer!"

William B. Leone

Born: 5/23/1947; Rochester, NY; 5 younger brothers.

Parents: Alfred and Lillian Leone; My dad was a WWII veteran. He served in an Army Construction Battalion.

Graduated: Aquinas Institute, St. John Fisher University; St. Bernard Seminary

Served Diocese of Rochester for 50 years with 12 assignments over the years.

Hobbies: Reading, golf, playing cards (Euchre), conversatons with people.

Favorite food: Spaghetti and Meatballs.

Lee Nelson

Lee was raised until the age of eleven in Rochester NY. His mother remarried in 1984, and this led him to Waterloo NY, nestled in the Finger Lakes region, and the proud birthplace of Memorial Day. He joined the Air Force in June of 1992 and completed one enlistment of four years. At the time he believed he required discipline and direction, and he got both. The military changed his life. He returned to upstate NY in 1998 and went to nursing school on his Montgomery GI Bill. Prior to nursing school, he was a surgical technologist and has now been a nurse since 2005. Altogether, military service included, he's been in the medical field for thirty years. He's presently an operating room nurse, and he and his loving wife Nicole have a deli, Lakeside Deli and More, in Irondequoit NY where they've lived happily and peacefully for fourteen years. In their blended family, they have three adult children.

In his spare time, Lee has always aspired to be a writer. He's presently putting together his first book of poetry and has also written several short stories, an incomplete novel, and one screenplay in desperate need of revision. His other hobbies include weight lifting, camping, landscaping, craft beer and The Bills Mafia.

Lee had to leave the military to love the military. This truth may reflect in his accounts. He's glad he did it, glad it's over, and honored to be among those who chose to serve. He thanks you for your time, and he never would've missed the opportunity to share.

Joe Mele

Joseph Mele Jr. retired from software test engineering in August of 2018. Since then, he's had time to discover the stories in the hundreds of letters that were found when his mom passed on in 2012. The letters date back to 1934. Many of them are from the WWII years when Joe's father was the surgeon with the 749th Tank Battalion fighting through France, Belgium and Germany. It has become Joe's mission to provide background and context for the letters, and to tie them into a cohesive narrative. The Rochester Veterans Writing group provides insight into a soldier's experience, as well as friendship, support, and healing. He considers it a privilege to participate actively in the group.

Noel Nichols

MM3 US Navy 1961-1965

USS Hopewell DD-681 62-64 WPAC Cruise

Tincan Sailor field Day USS Sullivans Naval Park, Buffalo, NY July 2001.

Sailors of the scrapped, sunken and lost fleet.

Gary Redlinski

After an education to pursue a career in funeral service as a director and embalmer, Gary Joined the U.S. Army in 1967 and trained for the Graves Registration MOS. From May of 1968 to July 1970, he served as an Identification Specialist at the Army Mortuary located on Tan Son Nhut Airbase, outside of Saigon (Ho Chi Minh City), Republic of South Vietnam. During his tour at the Mortuary, he handled and processed thousands of soldiers' remains. His writings reflect those memories of his experience as a soldier during that war.

Arash Rivazfar

I was born in Pensacola, Florida, and raised there by my single mother by the age of five before my father got custody of us. I lost my sister Sara to a terrible incident when I was five. We moved to Rochester, New York, where I grew up with my six sisters. I was the only boy and the second oldest of the six children. I was taught *family* was everything, and I was to watch over my sisters and protect them. I lettered in three sports in high school was a great student. School was first in my eyes. My freshmen year I knew I wanted to serve my country so I went to my father and asked him to help me sign up for the DEP program for the United States Navy. I left for boot camp July 29, 2002.

John Michael (Mike) Ryan

Mike was born in East Rochester, NY and resides there now (2024) with his wife, Juanita. They have a son and a daughter. Mike's childhood hobby was building models of Navy ships. Now, his pastimes are playing pool and cards. His current love is his 2004 Chrysler Crossfire. He is retired from a post-service career in printing. Mike is thankful for the lessons of independence and confidence building that he experienced in the Navy.

Sue Spitulnik

Sue Spitulnik was an Air Force wife, living in four states and England from 1972 to 1979, then worked until 1990 in a Tacoma, Washington restaurant where the clientele was military connected. "Immersed in the culture," she says. She moved back to her home state of New York in 1991, now residing in Fairport with her husband and two cats.

She has been a member of the RVWG since 2015 and had an integral part in their first two published books and is a happy contributor to this one. She is starting her seventh year (2025) as the president of the Lilac City Rochester Writers, a fiction writing group that focuses on showing writers how to improve their craft. In 2024 Sue chaired the Women Writing the West Conference at their live gathering in Denver, Colorado.

When Sue isn't organizing something or writing, she spends time in her quilting studio, with family, or out listening to live music.

John Steele

Master Sergeant John Steele served in the Army and Army Reserves from 1985 until his retirement in 2015. He entered the Army as a combat engineer, worked his way up to squad leader and platoon sergeant, and eventually worked in positions as an instructor and in administration. On the civilian side, Mr. Steele taught Special Education for eleven years, and after 9/11 worked for the Department of Defense for another decade. He started dabbling with writing in 2011 during a year-long tour in Iraq, and joined the Rochester Veterans' Writing Group in 2015.

Vaughn Stelzenmuller

Vaughn Stelzenmuller writes, because it's hard to stop doing it. His five-year military career piled up ample material to do just that, fortunately with plenty of humorous events along the way.

After commissioning into the Regular Army from Johns Hopkins University in 1967, the government posted him to Fort Hood, Texas; Central High-lands, Republic of Vietnam; and then Stuttgart, Germany, with the usual Army schools along the way. He writes about what happened as an infantry officer, and many of those things can be humorous, except sometimes to himself. When veterans write, everybody learns.

Vaughn made captain after two years' active service like all unlucky first lieutenants at the time, and also avoided getting sent to the stockade. After active duty, Eastman Kodak Company hired him as a manufacturing engineer. Now he writes, and still hopes for "promotion to captain" in real life. He gives it another decade or so.

Richard Terboss

Richard Terboss was born in the foothills of the Endless Mountains region north of Scranton, PA. After graduating from college, he came to East Rochester, NY in 1962 to work for Xerox Corporation. He worked for them for thirty-five years minus three years in the US Army Military Intelligence Corps. He still lives in East Rochester (2024) near his three daughters. His wife, Dorothy is deceased. His retirement hobbies are writing poetry and narratives and reading. He attends sporting events and belongs to two veterans groups. He loves choral and instrumental concerts and plays. Now in his four-score-plus years, he identifies with the saying that goes: Don't know the future but the past keeps getting clearer every day.

In *Important* Memoriam

Bob Whelan

Bob Whelan was a beloved member of the Rochester Veterans Writing Group and a great friend to all of us. His sudden absence leaves holes in our hearts as big as bomb craters.

On Saturday, November 7, 2020, we dedicated our monthly meeting to writing stories of our fondest memories of Bob and what his friendship has meant—and will always mean—to each of us. He touched each individual, as is plain from the stories that we see in this book.

We use "prompts" as a catalyst for our monthly writings. But somehow, prompts or no, these stories would have come forth anyway. Bob had that great an impact on us.

Rest in peace, dear friend. You gifted us with many stories that inspired us for "The Lighter Side of Military Life."

In *Important* Memoriam

Charles F. Willard

Then (August 1960) and in 2024

Chuck was another beloved member of our Rochester Veterans Writing group. He was raised in a rural setting, tending the family's animal herds from age six. He attended Syracuse University, studying engineering and enrolling in Air Force ROTC. This led to a four-year career as a pilot of C-130 Hercules aircraft, flying missions to 42 countries, including numerous operations in Vietnam and Southeast Asia. In his subsequent engineering and management career at Xerox, Chuck led the technical team that developed the world's first laser printer. In addition to all the stories Chuck has in our three books, he published Air Force Adventures first and second edition, and the Juanita Trilogy, an historical fiction series. His books are available on Amazon. We will always remember him, an "old-bold pilot" who just wanted to still be flying C-130s.

Made in the USA
Monee, IL
04 November 2026

67383302R00174